FIGHTING FOR YOUR LIFE

Real-life stories of heartbreak and hope, direct from the frontline of the NHS

FIGHTING FOR YOUR LIFE

A PARAMEDIC'S STORY

LYSA WALDER

WITH JACKY HYAMS

JOHN BLAKE

Published by John Blake Publishing,
80–1 Wimpole Street,
Marylebone
London W1G 9RE

www.facebook.com/Johnblakepub
twitter.com/johnblakepub

First published in paperback in 2008
This edition published in 2020

Paperback ISBN: 978-1-78946-204-3
ebook ISBN: 978-1-84358 234-2
Audio: 978-1-78946-255-5

British Library Cataloguing-in-Publication Data:

A catalogue record for this book is available from the British Library.

Design by www.envydesign.co.uk

Printed and bound in Great Britain by Clays Ltd, Elcograf S.p.A.

1 3 5 7 9 10 8 6 4 2

John Blake Publishing is an imprint of Bonnier Books UK
www.bonnierbooks.co.uk

This book is dedicated to the memories of Stephen Rowbury, Stephen Wright and Dorothy Wynter – wonderful friends who were sadly taken from us too young. They made a huge impression on everyone who knew them. You'll never be forgotten.

FOREWORD

If you live in the capital, you're aware of them all the time. They are the noisy ambulance sirens, the helicopters whirring above your home at 3am, the speeding police cars and clamouring fire engines racing through the traffic and down the clogged city streets to the next emergency, the next drama.

The sound and fury of London's emergency services might irritate or annoy the frustrated motorist, the pedestrian trying to talk into their mobile or the edgy sleeper, wakeful at any hour. But this background chorus, strident as it is, is an omnipresent soundtrack of inner-city life, as familiar a sound in London as the sight of red buses, nose to tail in Oxford Street, or black taxis crawling up and down Shaftesbury Avenue.

But what lies behind it all, behind the thousands of 999 calls made every day in London, home to 8.9 million people, a city that never sleeps? What happens when ambulance emergency crews turn up at an incident only to encounter injury, mayhem, danger, madness and, in a few cases, tragedy? Who are the people whose job involves walking down the city's streets, night and day, tending to the victims of crime or sudden illness? And what sort of dangers do they face as they go about the business of responding to the emergency calls?

Over the course of a four-year period, as a feature writer for the London *Evening Standard*, I interviewed many people working at all levels for our public services – and Lysa Walder was one of them. I talked to doctors, nurses, A&E staff, police officers and emergency teams. These are people whose daily toil involves dealing with the consequences of illness, alcohol, drug addiction, crime and disaster – sometimes facing the very extremes of human behaviour in an unpredictable working life where anything can happen, at any hour of the day or night, in and around London's teeming streets and suburbs.

These are the people that mop up the blood, pick up the pieces, soothe the scared or terrified and, on a good day, save lives that would surely have been lost without their input and speed of response.

I'm a fully paid-up cynic but it is no exaggeration to say these London emergency workers like Lysa are impressive in their dedication to the job, professional and yet modest about the crucial role they play in the life of the city. Lysa, like so many of her breed, insists that there's no time for individual heroics in the world of 999 and emergency services. Only groupings of fire, police, ambulance, helicopter and medical teams who know that the teamwork itself is what makes the service work efficiently, not individual acts. A nurse, paramedic, mum-of-three and Londoner, Lysa typifies to me what the ambulance service is all about – down-to-earth, professional, highly trained, caring people who get behind the wheel and get on with the job, no matter what it may involve. And in the combat zone that some of our meaner streets make up, they deploy fast response, humour and a skilled but very human approach to a job that most of us would baulk at.

Lysa's stories include that of the woman whose husband tried to slaughter her, the baby condemned to a life of abuse, the young life that lay bleeding, trickling away from a senseless knife wound on a south London street. Her voice comes directly from the frontline – and the heart. What she tells us about the day-to-day work of London's amazing free ambulance service – the biggest and busiest such service in the world – may be scary, shocking, sad, thought-provoking or downright funny but it also tells us a lot about human nature and ourselves, wherever we live.

But if Lysa's stories are a form of acknowledgement of the work of London's ambulance service, it must be said that recognition for the high standards and dedication to the job should be made equally to all NHS and St John Ambulance emergency crews up and down the country.

It's a tough job at times but you do us proud. You cope, unflinchingly, with fear, pain, shock, death and out-of-control behaviour, and put your own lives on the line. And you usually do it with a smile and a joke. Thank you, every single one of you. You insist you're not heroes. But to many of us, that's what you are.

Jacky Hyams, London, 2020

CONTENTS

INTRODUCTION

I've always liked a bit of adventure in my life. I guess that's why I ran off to join the circus as an incredibly naive sixteen-year-old, much to the dismay of my mother. But who could blame her? It's not every parent's dream career for their child. My father, however, had spent some time as a travelling musician, so he understood the appeal of life on the road.

During my four years in the circus, I performed a trapeze act, juggled, walked on a huge globe, rode a horse and had a straight role in a clown act. I even had a brief role as ring mistress in one show, mainly because I was the only one who spoke English without a strong accent! Working in the circus took me around the UK, Eire and Europe, and I managed not to break my neck – or anything else for that matter.

Aged twenty, it was time to leave the circus and study for some sort of career, I started to train as a nurse. What made me choose nursing? To be honest, I'm not sure I gave it much thought at all. I suspect I'd been watching a bit too much

Casualty on TV. Three months into my training at nursing college, I was finally let loose on the poor unsuspecting patients.

Finally, there I was on my first ever nursing shift, self-consciously wearing my folded card hat and swanky starched dress, raring to go and save a life or two. For my first task the ward sister directed me to help a staff nurse, who was already with an elderly man behind some curtains.

'Hi, I've been asked to help you,' I said brightly.

'Great,' said the staff nurse. 'I'll stand Mr Smith up. Can you wash his bottom?'

'Are you serious?' I wanted to scream in horror. It had never really occurred to me that I'd have to wash the heavily soiled bottom of a fully-grown man: I said I was naive. However, I dutifully got on with the task at hand. In my head, however, I was already planning my escape. I'd find some other work soon, I consoled myself. I'm not ever going to do that again!

But I never did get to hand in my notice because, despite my mucky introduction, I actually began to enjoy nursing. Towards the end of my training as a third-year student, I spent time in the Emergency Department (ED) and found 'nursing' as I'd always imagined it to be; busy, exciting and unpredictable. In the emergency setting, every day is different. Just my cup of tea.

During my stint in ED, I had to spend a day observing the work of the ambulance service. Some of the other women had already warned me about the ambulance men.

'They're a bunch of womanisers,' said one.

'They're all sexist pigs,' said another.

'Be careful or they'll have you running around making tea for them all day long.' Back then, although there were ambulance women, they were very much in the minority.

By the time I got to the ambulance station, where I was to undertake the observational shift, I was terrified. Acutely conscious of my nurse's uniform, I felt so nervous. Sitting in the mess room with all the other ambulance staff. I was almost too scared to speak. It was so different from the predominantly female environment I'd been used to at the hospital, and the conversation was far from politically correct at times.

One of the paramedics offered me a cup of tea. Is this a trick? I thought as I nervously said, 'Yes, please.' It wasn't, he just handed me a horribly strong cup of tea. Sitting there, nervously pulling my skirt over my knees as we waited for the first emergency call to come in, I burned my mouth on the hot tea as I sipped it. It really was too strong. But I was too nervous to say a word.

After a short while an emergency call came through. Two ambulance crew strode over. 'Are you coming with us then?' said Steve, the taller of the two. Silently grateful to leave the horrible cup of builder's tea behind, I joined them. Little did we know that day, but Steve was destined to be my husband (the poor, long-suffering fool!).

We made our way to the location of the incident, lights flashing, sirens blaring.

It was what we call 'a proper job'. A woman in cardiac arrest. Right there in the street, I watched as Steve, a paramedic, passed a tube into her throat to allow the oxygen to be pushed into her lungs. This is called intubation. Then he put a needle into her vein to provide a route for fluids and drugs to try to restart her heart, a process called cannulation. In the meantime, I tried to help a little by administering the cardiac, or heart, massage. Then we lifted the woman onto the trolley bed and moved her into the ambulance. A priority call went out to the local hospital and Nigel drove us there with the blue lights flashing and sirens blaring. In the back of the ambulance, Steve and I continued to resuscitate the woman during the journey to hospital. I hadn't known anything like this before. I was on an adrenalin high all the way.

When we got to the hospital, I was shaking like a kitten. Bits of my hair had dropped out from my ponytail; I was hot and sweating from the exertion of resuscitation during the six-mile journey. Even my stockings had ladders in them and I had grazes on my knees from where I'd kneeled on the pavement while performing cardiac massage. I looked a complete wreck, but I was buzzing.

To me, that job had been amazing. Sadly, as is often the case in such situations, the woman did not survive. However, I knew, there and then, that this was the type of challenge I enjoyed. This was the environment I really wanted to work in.

In 1993, I finished my nursing studies and became a

registered general nurse. Back in the early nineties, getting a job in the Emergency Department wasn't that easy. There were no vacancies. I had to put my name on a waiting list for the ED and in the meantime started working as a staff nurse on a general medical ward, hoping to get a chance to transfer to an emergency setting. Hedging my bets, I also quietly applied to join the ambulance service and within six months I was accepted. At last I'd be going back to life on the road. But this time it wasn't wearing a diamante bikini in a travelling circus. I'd be pulling on steel-toe capped boots in the London Ambulance Service.

I started working as an Emergency Medical Technician (EMT) in April 1994. EMTs formed the majority of the frontline staff, responding to all types of 999 calls and providing basic treatments for resuscitation and defibrillation, bringing arrhythmic heart contractions under control, and administering many emergency drugs and other treatments.

In 1996, I opted to train as a paramedic and, in 2003, I undertook additional university education to become one of London's first Emergency Care Practitioners (ECPs). ECPs responded to all types of 999 calls too, but in the London Ambulance Service they often worked in cars as single responders. Because all ECPs were also paramedics they were sent to high-priority calls, like cardiac arrest or trauma. An ECP in a car could manoeuvre through the traffic that bit easier and start treatment while waiting for the ambulance crew to arrive.

At the other end of the scale, an ECP was also sent to lower-priority calls, like minor injuries or illnesses. In these situations, many patients may not have actually needed to go to hospital. An ECP could carry out a full examination and assessment on the spot and frequently carry out the care or treatment that traditionally could only be provided in hospital. We carried additional testing equipment and could administer various medications, including antibiotics, on the spot.

Many ECPs also worked in minor-injury units, walk-in centres and EDs, as well as with the ambulance service. This meant we could work alongside other healthcare professionals and had the opportunity to learn more from them. The skills we learnt in these other settings could then be applied back out on the road. Elderly or housebound people in particular were often very grateful for this – because it meant some didn't need to go hospital at all. An ECP could also refer patients back to their GP, community team or other treatment centre for continuing care. That is why I found the work so rewarding: it was incredibly satisfying to be in a position to help someone and leave them at home, happy and smiling afterwards.

I left the London Ambulance Service in 2014. I now live in an old farmhouse high in the hills in northern Tuscany. I return regularly to work in Urgent Care Centres in South London as a Specialist Paramedic.

I wanted to write this book because there's no doubt that this kind of job seems to capture people's imagination. In a social setting, people do seem very keen to know more about our work. More importantly, I wanted my children to know about that important part of my life, to capture it for all time. I'd like to add here that any thoughts or feelings voiced in these stories are mine and mine alone. I'd never suggest that I represent all my colleagues in that respect. But it must be said that the stories you will read here are stories any one of us could tell. We've all been to sad, traumatic, scary, funny or ridiculous call-outs. I don't have any monopoly on that!

The people I worked with in the ambulance service were a fantastic bunch. I never laughed as loudly – or as heartily – as when I was sharing a joke in the mess room with my friends and colleagues. There was a real sense of fun between us and we socialised a lot. In fact, I believe there are very few work environments that shared our sense of camaraderie. Perhaps that's one reason why so many staff date or marry each other. I'm a testament to that, of course, as I married Steve. It isn't that unusual for ambulance staff to marry nurses, doctors or workers in other emergency services too. I guess that's because we all have so much in common, like dealing with unsocial shift patterns. And, of course, life and death.

TRAGIC

ADDICTS

A heroin overdose usually comes through to us as 'young male collapsed' or 'difficulty breathing' – a heroin overdose will stop you breathing. But when someone dials 999 to report such an incident, they're unlikely to want to flag up the truth. If they're heroin users, their main concern is not alerting the police.

Addicts buying heroin sold on the street don't really know the purity of what they're buying. It could be cut with other stuff. Or they might get the dose wrong. Seconds after injecting what they think is the right dose into their vein, they may stop breathing effectively or in fact altogether, while turning grey-purple, and their heart may stop beating through lack of oxygen.

There are times when a heroin overdose can be a very quick end. But in some cases the person's breathing just trails off. So if we are called there in time, there's a chance we can save them. In this situation we can give them a drug we carry at all times. It's called Narcan and it reverses the effect of any narcotic-based drugs like morphine or heroin. It can actually reverse the effect of an overdose within minutes.

But amazingly to us, heroin users are not always grateful or relieved if we manage to save them from a sad end. Often, once the effect is reversed by the Narcan, they're up, alert – and angry. Their view is they've spent their money on heroin and we've gone and wiped out all the 'benefits' as far as they're concerned! Amazing, isn't it? But we don't say, 'Excuse me, you were *dead* a minute ago and you're still worried about your drugs?' But that's really what they're thinking. Sometimes they'll just walk away from us.

This can be more dangerous than you'd think. Because the life of the heroin still in their bloodstream is longer than the life of the antidote, there's a chance they'll go into respiratory arrest again. If we can sense that they're likely to run off straight away – and you sometimes get to spot the signs – we may give them another injection of Narcan. We don't want them to collapse. For us, it's a case of making the best of a bad job. But only if they let us.

Drug addicts are occasionally verbally abusive. But over time you realise that the verbal abuse you might get from them isn't really personal. It's just aimed at whoever is doing the job. If you're female, the abuse is 'fucking blonde bitch' or 'stupid blonde cow'. If you wear glasses, you're a 'speccy git'. A bald paramedic (like my husband) is a 'bald git'. And so on. The abuse or aggression can be much worse when you're trying to help someone on crack. Heroin's a bit of a downer but there's no reasoning with someone on crack. You're wasting your time.

Tonight I'm called out with my husband Steve to a familiar address. It's a rundown bed-and-breakfast, a hideously depressing building, tatty, seedy, hasn't been decorated or touched for years. At the front desk there's a scruffy, unshaven, unkempt individual with a fag clamped firmly to his lips, resembling Onslow, the slob played by Geoffrey Hughes in *Keeping Up Appearances*. The whole place reeks of tobacco, stale sweat – and despair.

'Have you got a room for two weeks in August?' quips Steve, deploying gallows humour: we know we're probably going to need it. The call's come through as 'breathing difficulties' – but we're pretty sure someone's overdosed on heroin.

'I've never been called out to this place for anything else,' I remark to Steve as we follow Onslow down the shabby corridor to a squalid room. It reeks overwhelmingly of cannabis.

Sitting by the window is a twenty-something man slumped in a chair. He's not a pretty sight. His skin is mottled, purple. A crust of dried vomit is covering his face and chest. Three other people, two men and a girl, are just standing there blankly, no flicker of any kind of emotion from anyone. They're completely spaced out.

One of the men gestures to the chair. 'He's not very well,' he says in a matter-of-fact way. 'We had to resuscitate him.'

'When did he last talk to you?' I ask, putting down my bag.

'Oh, just a few minutes ago.'

Steve and I start to move the man from the chair to the floor so that we can get going on resuscitating him. But it's impossible to move him. He's as stiff as a board. It's almost as if he's moulded to the chair. Now I'm confused. Stiff means rigor mortis, when all the muscles in the body become stiff and inflexible. It's a good indication of death, because you find it happening in the body in the first two hours after death – and after about 8–12 hours the body becomes completely stiff. So this man's been dead for hours. Not minutes.

'Are you completely sure he was talking to you a few minutes back?' says Steve.

'Yeah, he was movin' around an' everything.'

Unlikely.

As paramedics, Steve and I know there's no way in the world that this is true. The man is very, very dead. And he has probably been so for some time.

Now another paramedic turns up, ready to help us. But I stop him. There's nothing at all we can do. Again, we try to get some sense from the trio: what happened?

This time the story takes a slightly different tack.

'Well, we were up all night 'cos we were talking. And at about 6am we noticed he'd stopped breathing. But we did resus and we got him back, we did. And then we put him in the chair. He was movin', definitely. So then we went out for a bacon butty from the cafe.'

Nice, eh? But whatever you might think of their behaviour – leaving a half-dead friend because you suddenly decide you want your breakfast – in these people's minds they actually think they did a good job. We don't believe for a minute that they really did make any attempt to revive him. What they probably did was reluctantly call 999 when they got back from the cafe. And in fact he'd been dead for hours.

The police were called but it wound up as 'a suspicious and unexpected' death and we were never asked to give evidence. Probably it was an accidental overdose. But, unluckily for him, he was using in the wrong company. They just didn't realise he was gone. But that's not unusual. A lot of people don't know a dead person if they see one.

On TV or in the movies, death is always violent, there must be blood coming from somewhere. Fictional death always has a lot of drama around it. Whereas in reality most people die fairly intact. And look quite peaceful.

Quite often you'll go to a house and someone will say, 'She's upstairs, dear. She's been very quiet.'

And very dead.

ATTACK IN THE PARK

We're in the ambulance station and I'm bored as hell. Some people like it when it's quiet. Quiet times mean sitting around, joking, watching TV, gossiping. But it's too flat for me with nothing happening. I prefer to be busy with challenges, things happening fast.

'I want a really interesting job right now,' I say to Carole. 'Something different. Haven't had one of those for ages.'

Carole looks at me as if to say, 'Are you bonkers?' She's one of the quiet life brigade.

Don't they always say, 'Be careful what you wish for'? Out of nowhere, we get what's termed 'an abandoned call'. It means someone has rung 999, given a bit of detail and just hung up: the only information is a man and a woman have been attacked in a south London park, no precise location. The police are already on their way. Can we meet them at one of the entrances?

By the time we arrive, the police have had more 999 calls from the man who has been attacked, ringing on his mobile.

He's been trying to explain to the police where he is in the park, but they can't yet locate him. He keeps saying he's near the bandstand – which doesn't really help. He's told them he doesn't have a clue where his girlfriend is. The story is, they've both been attacked by a gang of black men, but the men took his girlfriend away.

The best option for us is to get to the closest park entrance to the bandstand. We get in there with our equipment and torches. Two of us, me and Carole, go with police officers into the dark and deserted park, shining our torches around. But we get nowhere: not a sign of anything. Back to the vehicle for another police update.

And now a bit more information comes through: the girl is just 16. Police have asked the man if he knows if she is still breathing – he can't confirm either way. He keeps telling them she's been beaten up and taken away.

Now it's getting quite serious. We've already lost 20 minutes looking for the girl and not only is there no trace of her, we don't know what state she's in, how bad the attack was. Then a breakthrough: one police officer has located her and directs the control room to 'the middle of the park'. She's not breathing, he says, and has no pulse. He tells control he's already doing mouth-to-mouth on her. Overhead a helicopter buzzes, joining in the search for the elusive 'gang'. A group of kids on bikes are relishing the kerfuffle.

'What's going on, who's dead?' they hassle us.

'Go home,' I tell them. 'It's very late.'

We're frustrated, to say the least. As the minutes tick by, this young girl's life could be fading away. The sooner we get to work on her, the better. Now police are directing us to a wooded area. It's virtually inaccessible: to get in we have to climb over a six-foot-high metal fence. And finally, in a clearing, we see the lone copper, desperately carrying out mouth-to-mouth resus on the girl.

He looks up at us. 'I found her like this,' he says. We tell him to keep going while we set up our equipment. Despite what the public may imagine, one thing 999 crews don't do as a rule is mouth-to-mouth, unless it's a real emergency with a baby or small child. There's a risk of contamination or infection via saliva so it's regarded as too dangerous for ambulance crews. But the policeman quite rightly wants to do something, anything, until we turn up.

Carole and I then start to use the bag and mask to try to get oxygen into her, ventilate the girl's lungs. I notice purple bruise marks on her neck and this important bit of information is instantly relayed to the police: could be attempted strangulation. Then Carole starts chest compressions on her, irritated by the helicopter whirring noisily above, shining the spotlight right on us. 'Interesting enough for you?' she snaps at the chopper. We're desperate to get the girl back. And yes, after ten minutes, with four of us working away, we get a sign of life. Her pulse comes back. And her heart is starting to beat for itself. It means there's hope at least. But she's still not breathing properly for herself, so we continue to use the bag and tube to squeeze oxygen into her lungs. Another ambulance has turned up to help us. And the fire brigade are trying to cut open the gates to the area.

While we've been working, a group of police have finally managed to locate the boyfriend elsewhere in the park. 'Do you want a quick look at him?' we're asked over the radio. We refuse. The girl is our priority. And the police say there's no obvious sign of any attack on the man. They'll get another ambulance for back-up.

Finally we get the girl out of the park and on the way to hospital. She's still got a pulse as they wheel her into intensive care. But the problem is, we spent too many crucial minutes trying to find her. So there could be brain damage – that's if she survives at all.

The hospital is swarming with police. They're trying to find the girl's family. The man has been taken to another hospital, accompanied by police officers. Already they are starting to have serious doubts about his story. For starters there are the marks on the girl's neck. And their suspicions only increase with the boyfriend's detailed description of the attackers' clothing. The fact that he could recall that they all wore Nike trainers, for instance, strikes a very odd note. The last thing anyone recalls when they're being attacked is the brand of shoes or top their attacker is wearing. And, though he claims he's been beaten up, there's not a single piece of physical evidence to back this up.

For a few days it's touch and go for the girl in intensive care. Tragically, despite everyone's best efforts, she dies, surrounded by her family, who keep a round-the-clock vigil in muted shock. Not long after, the police have enough evidence to charge the boyfriend with her murder. They've had real suspicions right from the start, but they let him continue with his story – and he keeps stringing himself up with every detail. He admits that he's taken the girl for a romantic stroll in the park – and they've had sex: he insists it was with her consent. Then, he insists, the 'gang' attacked them out of the blue.

No one knows what really went on between the pair that night in the park. But the evidence shows he strangled her. And fortunately two very brave ex-girlfriends eventually stand up in court and describe how he'd attempted to strangle them – after they'd tried to finish with him. He is subsequently convicted of murder. It looked as if this girl, too, had tried to end the relationship. But tragically she died for her wish to end it.

So at least justice has been done. And yes, this was a challenging job. I'd have preferred a happier ending.

GETTING BY

I can hardly believe what I'm seeing. I'm in Norwood, trying to get into an old, dirty and unloved Victorian house that looks like it probably saw better days a century or so back. I struggle even to get the front door open; the owner has left the key hanging from a shoelace inside the letter box, but I have to fiddle around for ages to locate it. When Rick and I finally manage to get in, it's quite a shock. 'Squalor' just about describes it. Think *Life of Grime* on TV and you've got the idea.

First there's the smell. It's overwhelming, and knocks you back as you walk in: a combination of cat's and human urine. Paint is peeling from the walls. And every single inch of floor space that the eye can see is covered with rubbish, boxes, newspapers, cereal packets, paper bags, envelopes, plastic bags, every bit of junk mail that came through the door – nothing has been thrown out of this house for decades. Someone's been living like this for years.

The kitchen would have Aggie and Kim from *How Clean Is Your House?* in raptures. They could do a whole series just on this one home. Every surface is thick with grime and grease, filthy plates, smelly rubbish and empty tins everywhere. Health, safety, fire hazard, you name it – this place is a rotting tip. And there are manky cats running around. Somehow they're managing to survive in all this filth.

The control room say there is a man in his seventies with chest pain and difficulty breathing. The control room warned that we'd have to let ourselves in. He's in a ground-floor bedroom, surrounded by another tidal wave of rubbish, junk and yet more paper, an old man, lying there in urine-stained, wet, filthy sheets. By the bed are a couple of dirty glasses containing some sort of liquid pond life. He's in his underpants – and very short of breath. Somehow, in all the dirt and chaos, he's struggled to get by all these years. But now illness has taken over and he's stranded in his home, adrift in a sea of grime.

'I can't breathe,' he groans at us. Rick and I manage to sit him up but there's no question: he's very poorly and needs the hospital. But it's quite difficult for us to get to him; clambering around the boxes and clutter to reach him is no easy matter. How on earth did he even manage to get around? He must have been crawling over all this stuff.

I try kneeling on the bed – and note that the urine is already seeping through my trousers – but it's impossible to stand properly on the floor, it's so thick with clutter.

'We're going to get you to hospital,' I tell him. I sound like I'm in control, but I'm not because I can see there's a real logistical problem involved in getting him out of here. There's no way this man can walk. And there's no way we could get our carrychair, a version of a wheelchair, out to the hall or the front door because the rubbish and clutter everywhere make it impossible to do this safely. So we have to call control and ask for assistance from the fire service. We manage to set up an oxygen mask to help him breathe. And then we get a bit of his story. His wife died many years ago and he hasn't been able to get out very much.

'My neighbours help me,' he tells us.

'But does anyone else come in to help you?'

'No,' he wheezes. 'I don't go to the doctor and I don't take medication. I don't want anyone coming here. I can manage.'

How many times have I heard those three little words? Frail, elderly people living alone who want to cling to their independence say that to us time and time again – when the evidence is so clearly the opposite. They say it to concerned relatives too, usually the ones that ring up but, for whatever reason, usually distance, never actually get to see them. So it goes on and on for years. Technically they're 'managing', 'getting by'. In reality, it's all out of control. But this really is the worst chaos I've seen for a long time.

Luckily the fire service have turned up so we can organise his departure. It's a real struggle but four of us manage to strap him on to a board and we form a small chain so we can easily pass him along to each other. The fire service have decided our best bet is to get him out via a sash window in the bedroom which I manage to open with some difficulty; it hasn't been opened for years. Outside there are more firefighters to help carry him into the ambulance.

'What about my cats?' he asks me. 'Who'll feed them?'

I suggest I ask the next-door neighbour who might be willing to put some food out for them. He nods. 'Yes, but as long as they put it on the front doorstep,' he warns me. Clearly he doesn't want even the neighbours to step inside. So before we drive off, I jump out to knock at the house next door. The neighbour, a pleasant-faced, plump, blonde woman, fills me in. They've been worried about him for quite some time. In fact they'd called the ambulance; they'd seen lights on in the house but hadn't heard anything of him for several days.

'He won't have a phone,' she tells me. But they're fine about the cats. 'He won't let anyone help him. Apart from us, he's got no one,' the woman says sadly. But they'd worked out a little routine: they'd been regularly collecting basic foodstuffs for him for ages.

At the hospital he is taken to the emergency department. Part of our work is to alert hospital staff to his living conditions. The fire service also report the house as a fire hazard. In due course social services will get involved and liaise with environmental services to clear the house. But I keep wondering about him. A few weeks later I manage to pop into the ward. A nurse told me the house has been cleared – and he's gone home. Apparently he's accepting a bit of help from social services too. So while it was probably the worst example of domestic neglect I'd ever seen, at least it didn't all end in total disaster.

Sadly our job involves seeing quite a lot of this kind of

neglect. Often the call-out itself is for a relatively minor thing. But when you talk to the elderly person you discover that they might only see another human being for an hour or two once a week. And the rest of the time they just sit there, in a chair, often in silence. If you're very old with impaired vision or hearing, even the simple pleasures of life, such as listening to the radio or reading, might be impossible. If you don't go out at all, no one visits, you can't walk much, you struggle to heat a can of soup or make a sandwich, what is left? Life just drizzles away from you. Yes, you're still alive. If you can see, you've got the TV. But in another way you're not really living, are you?

Yet it's not the dirt and the chaos that are the problem. The heart of it all is sheer loneliness. You can live alone in a spotless environment with everything you need around you. But you can still be totally isolated. I don't know the answer to that kind of loneliness, I really don't.

AN UPRIGHT MAN

He was 26. He died where he stood, absolutely upright, his body leaning against the bathroom wall. I've seen hundreds of dead bodies. But not many standing up like this, almost as if he's got a broom stuck up the back of him. It's a ghastly picture. His lower limbs are a mottled, purple colour because of the position he's died in and the natural effect of gravity. There's a name for it: it's called post-mortem staining. His face, shoulders and upper body are completely devoid of any colour.

I go through the motions, put a heart monitor on – flat line. Nothing. He's gone.

This is one of the dirtiest, most squalid council flats I've ever seen on any estate. Blood and excrement on the walls, floors sticky with heaven knows what. Disgusting. Empty cans of Stella everywhere, empty vodka bottles and ciggy butts wherever you look. Tobacco-coloured walls. The home of someone who's given up – and does nothing but drink, day in, day out. It's also cold and dark. The electricity's been cut off, making it even more sinister and murky.

'Is this a natural end?' the copper asks me.

Not really. This call had came through to me an hour before as 'male collapsed behind locked doors'. It's quite a common call-out and usually comes after a home help turns up and can't get any response from an older person. Tonight there were a couple of neighbours waiting for me outside the block. They called 999 because they know this guy. They led me to his flat on the ground floor. One neighbour had walked past this morning and noticed the man's head resting against the opaque glass of the bathroom window. When she came back in the afternoon the head was still there. 'He's a heavy drinker, he doesn't work – but he's not a troublemaker,' she tells me.

'Yeah, but you do get a lot of ambulance and police cars turning up,' chips in the other. I'm getting the picture.

Call-outs involving drunks are routine for any paramedic. Crews may even know the person, sometimes by first name. But I've never been called out to this guy. First, I bang really hard on the glass to rouse him. Nothing. I try again: bang bang. No response.

Now the neighbours are getting the message. 'Oh no, is he dead, dear?'

'I don't know,' I tell them – but I think I do. If he's been there since morning and my banging doesn't make him stir, it doesn't look good. But I can't get into the flat. I've arrived before the police but, as hard as I try, I can't manage to kick the door in. Getting in windows is easy, but breaking down doors is something else. I try and try – but I keep bouncing off it.

I try a final, hip-busting kick – but it still doesn't work. Technically we're not supposed to try to kick doors in – that's the police's job. When you think it's a worst-case scenario, though, you have some justification. You just have to remember that for every door that gets kicked in in London – and there are many – someone has to pay for it. So it's not something a paramedic does willy-nilly. Later we will discover five different bolts in the door. Clearly guests weren't welcome, though a ground-floor flat in a pretty rough estate is unlikely to be as safe as Fort Knox whatever you do.

Just as I'm starting to get really frustrated, three police turn up. One young officer starts to try to kick the door down in his size 12s. The door still won't budge. Another bigger one has a go – door kicking's a real macho sport – but nothing. On cue, a man suddenly comes out of another ground-floor flat, locks his door and – magic – he's carrying a tool kit. 'Excuse me, mate, but you wouldn't happen to have a crowbar in that bag, would you?' asks one copper. Hmm. The response isn't enthusiastic, though it's quite obvious what we're trying to do.

'Yeah, but I need it for work,' he grumbles.

'Never mind that, we need to get into this flat,' says the copper tartly.

Reluctantly the man opens his bag and hands over the crowbar, standing there wordless as the copper manages to jemmy the door open. He isn't the least bit curious or concerned about the fate of his neighbour. He just wants his crowbar back.

By now an ambulance crew has arrived. And inside the flat, once we've established that the occupant is dead, we try to figure out what has happened.

There's blood everywhere you look. It looks like he's stumbled around the place, steadying himself on the walls and doorframes. There's bloody hand and fingerprints – even footprints of blood on the floor. It's baffling because at first glance he doesn't appear to have any injuries. The police are thinking it might be a suspicious death, so we don't attempt to move him about too much. If it's a potential crime scene and he's been murdered, we'll have to remove our footwear for forensic elimination purposes and if anyone in the ambulance crew touches anything at all, police will need fingerprint checks from us.

When we start checking out the lounge we find the answer. It's a broken glass. It's lying by the sofa – which is covered with his blood. Somehow the glass has cut his arm and severed an artery or vein. An accident, it seems. And sure enough, when we check and look at the man again, we find the cut – which is actually quite small. But it was lethal. It looked as if he was sitting there, on the sofa, for quite some time, just dripping blood and the sofa absorbed it like a sponge. He was in a badly inebriated state – the cans and empty bottles bear witness to that – and he must surely have realised he was bleeding badly at some point. It looks like he's staggered around the place, dripping blood everywhere, too far gone to do anything to help himself – and very slowly he's bled to death.

The police finally concluded that it was an accidental death. No one else was involved. He was definitely an alcoholic – and this was the tragic outcome, a lonely death in a setting that would give most people nightmares.

Later that evening, as I'm about to finish my shift, I run into a friend, another paramedic who works in the same area. I tell her about this young alcoholic who died at home and as I start to tell her about it, she stops me.

'Oh no, that's Gary. I know him. I got called out to him two weeks ago.' She's been called out to Gary quite a few times. She says that although he was always in a very drunken state, he was never aggressive or awkward and she'd often find herself chatting to him.

From what he's told my friend, life had been fine until he was 20, when he'd been severely beaten up in the street by strangers. He'd been left with a head injury. And that was when the drinking spiralled completely out of control. He'd told her he'd dreamed of being a songwriter. And he'd even sung her a couple of his songs while they waited for an ambulance.

'He was better than anything you hear on *The X Factor*,' she tells me sadly.

In this job you get used to dealing with alcoholics; some will dial 999 three, four times in one day – and invariably they refuse to let us take them to hospital. Over the years you watch people like this get worse and worse as they carry on with their slow suicide – we just do what we can, as do all the hospital staff. It's a slow means of destroying yourself. The alcohol blots out the reality – because, for whatever reason, the reality's too hard to bear. But when you think about a young man in his twenties who dreamed his dreams, had his hopes, just like we all do, well, it does stop you in your tracks, doesn't it?

EVE'S STORY

Oh no, this sounds bad: 'Female assaulted, multiple stab wounds.'

Instinct alerts me straight away that this call-out is most likely to be a really bad domestic. Women in London worry all the time about being attacked by a total stranger. But in fact 999 calls reporting an assault on a woman are usually due to attacks by the husband or partner. Or an ex. Sadly, despite all the awareness campaigns nowadays, domestic violence still manages to wreak its ugly havoc in too many lives. 'OK,' I sigh to Jim, 'on go the rubber gloves – we're gonna need them.'

As we pull into the street, police are swarming everywhere, waving us along to a house halfway down the road. In the front garden there's a small group of plainclothes guys too. They briefly fill us in before we can go into the house. The woman's husband has set about her with a meat cleaver in their kitchen. As she's attempted to get away, he's struck her many times, driving the cleaver deep into her back, her head and upper arms. A bloodbath. But she's survived.

We're steered through to the front room, but in the hallway we pass the kitchen and I see the husband sitting at a table – handcuffed, naked, looking terrified. He's shivering, surrounded by police officers but, despite what he's just done, I feel a momentary pang of sympathy. He looks vulnerable and confused. Is he mentally ill? I wonder.

Then we get to the poor woman in the front room. She's lying naked on her tummy underneath a quilt that one of the officers has placed over her. She's conscious, managing to talk to one of the officers. She's not even crying. But I'm horrified by what I see. It's a ghastly, blood-spattered horrorfest. He's lashed out at her over and over again all over the top half of her body, her head, her shoulders, upper arms. She's probably twenty wounds: some are minor, just glancing blows, others are really deep gashes, several inches long, on the back of her head. You can see the white skull glistening through one gaping wound. And they've bled so much you can't really see the colour of her skin, it's so saturated with blood. It's a shocking sight: Jim obviously thinks so too – and we're used to looking at things that would make most people throw up. This is starting to make me feel upset. How can a man do this to the woman he lives with and presumably loves?

In fact I'm not very far from tears as I bend down to talk to Eve. 'Are you in any pain, anywhere?' I ask her gently.

'No,' she says, then adds pleadingly, 'Please tell me what's happened to Joe. Is he OK?'

'He's fine,' I tell her because the police have already told me about Joe, her nine-year-old son. He heard the noise below but remained upstairs throughout the attack: he thought it was a burglar in the house. So he called 999, not realising it was his father who was the assailant. But in all the chaos, no one thought to tell this woman that Joe was safe and uninjured. A typical mum, she's thinking of her kid first.

Jim and I set about doing what we can. We manage to wipe her eyes, nose and mouth clear of blood and get some huge dressings over the wounds. We wrap her in a clean sheet, blanket on top, and carry her out to the ambulance. The police have closed the kitchen door so that husband and wife can't see each other – or the horror that his crazy attack has inflicted. We race to hospital with blue lights and sirens on and there the doctors and nurses quickly take over. Eve is taken to theatre for her many wounds to be stitched. It will take hours for someone to sew her up.

Our bit is over, but this is not the sort of job you'd forget in a hurry. Jim and I debrief in the ambulance afterwards, a little bit of therapy for us. We need it. We agree she's had a narrow escape. By sheer good luck, police had got there within a few minutes of the son's call. 'Supposing the kid hadn't been around, Jim, how bad would that have been for her?'

'Yeah, and it's lucky the boy didn't see it – or get hacked to bits himself,' Jim reminds me.

Later that day I'm back at the hospital and I track Eve down in the ward. She's sitting up in bed, swathed in bandages, looking a bit like an Egyptian mummy. All you can see are her eyes, nose and mouth. Even her ears have been badly cut and are hidden under heavy dressings.

I introduce myself and we start talking. In my mind, I'd assumed she was a long-term victim of domestic violence. Had her husband ever been violent before? I asked. I will never forget her answer.

'Darling, my husband is the kindest and sweetest man you could ever know. He's never hurt or even threatened to hurt either of us before.'

'But what happened today?' I say, not sure if I can quite believe what she's telling me.

'I just don't know,' she says sadly. 'It's not like him to behave this way – something's gone really wrong with him. I don't know what's wrong. But I'm worried.'

How awful. Instead of being worried for herself, she's concerned about him. Then she fills me in on the full story.

Eve woke up quite early that morning to the sound of what she thought was her husband just pottering and clattering around in the kitchen. But he was being exceptionally noisy, so she went downstairs to take a quick look at what was he was doing. She found him there, a loving husband of 12 years, stark naked, eyes glazed, pots and pans in disarray all around – holding their meat cleaver. Sensing immediately that something was dreadfully wrong, she turned to flee. But then he set on her, flailing around with the cleaver, hacking her again and again. Screaming her head off, she somehow managed to get herself into their front room and shut the door to keep him out. Meanwhile their son, Joe, heard the racket and sensibly phoned for help and he was spared the sight of the bloodbath downstairs. By then the husband had got into the front room by forcing the door open. He started to try to slash her again but the police turned up quickly and managed to restrain him.

I don't forget about Eve's story. And several months later, while parked in the ambulance outside the Emergency Department, a woman approaches me. I have no idea who she is until she gives me her name. It's Eve, virtually unrecognisable from when I last saw her, in her dressings and bandages. She looks perfectly normal now. Was her husband now in prison or a psychiatric hospital? Please don't let him be out and about.

'Darling, he's dead,' she tells me. I'm stunned.

'He had a brain tumour,' she explains. 'They only found out when they took him in for tests.' It turns out the awful attack on Eve was the first real sign that the man had a serious problem. After that his condition deteriorated very quickly – he'd died within weeks of attacking her.

I hug Eve and tell her how sorry I am. You can see she's quite brave, this woman, but life has dealt her a bitter blow. To lose your husband unexpectedly is bad enough, even if he hasn't nearly tried to kill you before. But to know that your child might easily have lost both parents if events had gone the wrong way, must be truly devastating.

OFFICE BLOCK

My first really serious incident after finishing paramedic training is at a tall office block in the centre of town. Two firefighters are already on the case, up on the canopy which hangs over the building's front entrance. All we know is that a man has jumped – or been pushed – from an office window. Somehow he's managed to land on the canopy. No information about which floor he fell from – or why he may have fallen. At this stage no one even knows who he is.

We reach the first floor and, lugging our equipment, climb out of the window to get ourselves on to the canopy. The man is lying on his back.

'Looks like it's all over,' comments my colleague Keith. The two firefighters are leaning over the man, furiously working his chest, carrying out basic CPR (cardiopulmonary resuscitation) to try to restart the heart, without success. They hear Keith, but say nothing. I look around: no sign of any blood at all. All very odd. Wordlessly Keith and I take over from the firefighters.

Now I'm starting to compress the man's chest. But I can tell that something is very wrong. 'His ribs are all broken,' I say. 'They're not offering any resistance when I push.' Then, kneeling at the man's head, I start to try to get some oxygen into his lungs. As I begin to squeeze the bag, blood spurts out of his ears all over my trousers. This is bad.

'Why are we doing all this? It's futile,' I hiss at Keith. He's a senior paramedic, knows the score.

He shrugs. 'Until we know a bit more, it's best to keep going, Lysa.' He's right. We don't know any more than what we see. We carry on with our hopeless charade. But now I'm starting to feel we're part of something bigger, a spectacle, a live show for the public. Because anyone round here who has time to stop and stare is doing just that.

We're quite near a bus shelter. A young dad lifts his small son up on top of the shelter, so he can get a better look at the paramedics trying to resuscitate the poor man. What a sight for a young child to witness. OK, it's human nature to have a quick look. But giving your kid a front-row seat to watch an unfortunate person fighting a losing battle for life is going too far. To me, it's despicable. How would these people feel if it was their colleague or partner lying there? Would they still stand there staring?

But now, almost on cue, things are happening: the welcome and distinctive whirring sound of a chopper overhead. The highly skilled pilots from HEMS (Helicopter Emergency

Service) can land anywhere, practically on a postage stamp. They've managed to land on the nearby road junction, much to the delight of the enthralled crowd.

Now the team are up on the canopy with us: two pilots, a paramedic and a doctor. They wear orange boiler suits. The doctor takes control. He examines the man, pronounces him dead. Within minutes they're gone, off to another call. Now it's left to us to move the body.

As we start to move him the full extent of the damage his body has suffered becomes obvious. His arms and legs move in a most unnatural way. It's horrible. He's like a human bean bag in the shape of a man. He just flops about. What has happened is, his skin has remained intact. But inside his body every single bone has been shattered to smithereens. We get him on to the board, strap his body, cover him with a blanket. Then the fire brigade help us get him on to a ladder and lower him down to the street and into the waiting ambulance.

Only now do the crowd start to disperse. It's a nasty job, this call, my first major trauma. I'm shocked at the prurience of the crowd, the state of his shattered corpse, my naivety that somehow he wasn't badly damaged because there was no blood. Then we get the gen from the police. The man was just 30. He worked in the building in a clerical job. He threw himself off the roof – some 17 floors up. They've managed to find some of his belongings up there, along with a suicide note. It's a first for me, the kind of first you could easily live without. I'm used to overdoses, suicide by hanging – but not the more desperate, violent extremes someone will go to in order to end it all. Despite my nursing experience, my time working in A&E (now ED – Emergency Department), I'm still shaken up by this episode. The 'bean bag body' keeps coming back to me for ages. How bad was it for him, I wonder, to do it this way?

The next day my sister-in-law calls me. She works in the building directly opposite. She and her workmates had actually seen something fly down from the roof of the building facing them. Then they watched us all working away – and realised it was a person they'd seen falling. She'd spotted me.

'I knew it was you, Lysa, because of your ponytail. But we couldn't understand why you were trying to resuscitate someone who'd dropped from such a height. I nearly rang you on the mobile to say, "Don't bother."' We chat some more. We both agree that people's ghoulish fascination with the worst – like the man who put his kid on to the bus shelter roof – doesn't say much for the sensitivity or compassion of strangers when another person's life is at stake. Then I remember another thing that galled: a Peeping Tom in an office building opposite, staring at us through a pair of binoculars.

'You live and learn about people,' I tell her. 'I know it's part of what I'm doing. But I'm not sure it's something I want to think about too much – or it'll put me off this job for good.'

As you can see, it hasn't. But you never forget.

THE FIRST DAY IN LONDON

Gerry has been doing the job for years – and he's quite jaded. He's seen it all, he says. Today we're called to Brixton: a man, 21, can't walk, legs hurt.

Blasé as ever, Gerry shrugs. 'This'll be a load of rubbish,' he grunts. He thinks it's a waste of time. I laugh. You never know.

Inside the big house, converted into many flats, two young student types, a girl and a boy, accompany us up the stairs. They're Italian. It's their flatmate, up on the third floor. He's been in bed for nearly twenty-four hours. 'He come home sick from work and go to bed,' we're told. That's all they know.

The bedroom is pitch-black, heavy curtains blocking all light. We can just about see the man because the sheets are pulled up right over him. They turn the lights on. I pull the cover down a little bit to get a look at him.

I don't expect this. He's got a horrible rash covering his face. It's purple, flat and blotchy, the sort of rash that doesn't blanch when pressed. It's deadly, a killer. Very bad news indeed. Gerry has a look with me. Cynic he may be, but today he's taken aback.

'Bloody hell,' he mutters.

We know what this is. It's serious blood poisoning, or meningococcal septicaemia, to give it its medical name. A lethal bacteria. For all the rash on the surface of his skin, he is bleeding just as badly inside.

Now we're taking a much closer look at the rest of him. The purple rash covers his entire body from head to toe. He's awake, coughing, confused.

'How're you feeling?' I ask.

'My foot hurt,' he says in heavily accented English. 'They hurt so much, I not walk.' Gerry immediately goes down for the carrychair so we can get him out and down to the ambulance.

His flatmates are stunned when they see his rash. Apparently he'd been working in a restaurant as a chef. It was his first day at work in London – he'd only arrived in the UK the day before from his home in Italy. He hadn't even been able to finish his first shift in the kitchen. They'd sent him home late evening as he was so obviously unwell. As bad as it is, I'm surprised he's not worse. The lower leg pain he's experiencing is associated with poor circulation. But you can also expect vomiting, dizziness and painful headache with this type of bacteria. It's terribly sad. He's so handsome, this young Italian, it's devastating to think of what's happening to his body. He's got thick, dark, curly hair and chiselled cheekbones, a real heartthrob.

Gerry and I don't discuss it – there's no need to – because we both know that his chances of survival are slim. At best, someone with this blood infection could face amputation of their hands or feet, because their circulation is so damaged. I feel so bad for him – but he's got no idea this rash is running rampant all over him. There's no point in mentioning it. It won't help. Given the likely outcome, all we can do is be as kind and gentle as we can.

We go to lift him into the chair. He's in real pain, every part of his body hurts. His friends watch in silence as we carry him into the ambulance. Now his blood pressure is dangerously low: his body is totally in shock. All we can do to help is radio through for a doctor to meet us, elevate his feet, to improve his circulation – and put a line into his arm. As we whiz through the streets I cling to a tiny, desperate, shred of hope: once we're at the hospital they'll be giving him the antibiotics he needs to pull him through, I tell myself. It's a short journey. Hope and despair run in turn through your mind in those few brief minutes. And there's a sense of helplessness too.

Finally we pull up. The doctor's opening the door. He looks at the young man, purses his lips, says, 'Yes, fair enough. Bring him in,' and the hospital take over. Gerry is strangely silent. We just get on with it and write the job up. There's another call within a few minutes: just a sprained knee, a brief distraction to take our minds off the tragedy we've just witnessed.

But an hour later the control room calls me. 'Your young man died,' I'm told.

I've been hoping in vain. How did he pick it up? No one would know. It's a bacteria that's normally passed on by someone carrying it if they sneeze or cough without covering their mouth. Or it can be passed on by kissing. We call it 'pillow contact'. He could have picked it up anywhere. And because he'd been coughing while we were helping him, Gerry and I have been exposed to the bacteria too. So we have to take a powerful antibiotic called pivampicillin, often given to people who come into close contact with a person with meningococcal septicaemia. It reduces the chances of developing the disease. Within a couple of hours of taking it, all your secretions are bright orange. Blow your nose, pee or cry – everything's bright orange for a few days.

I did shed a few tears for that young Italian. When you've sons of a similar age, it's all too easy to imagine his mother, hugging her son goodbye as he leaves home, suitcase in hand, full of happy anticipation, looking forward to his big adventure in a strange city. All her hopes and fears for him would have gone through her mind in that last farewell. And then, 48 hours later, she gets that dreadful phone call to say he's gone: every parent's nightmare.

This happened some time ago, before ambulance teams carried certain types of antibiotics with them in order to treat people on the spot. Nowadays paramedics can administer the drugs there and then before rushing a patient to hospital: a lifesaving development which means the public can get the right emergency treatment when they need it most.

ON THE LINE

It's impossible to get to every job on time on New Year's Eve. We do our best. But, for those few hours after midnight, it just goes mad. One particular New Year's Eve call will always haunt me.

I'm working with Amanda, with whom I've got quite a lot in common. For a start, our partners are both paramedics, not unusual in our line of business, given the combination of shift patterns and the nature of the work. We're just ready to start 'greening up' (showing a green light on your console means you're available for the next call) at the local hospital. It's coming right up to midnight, so there's a big group of ambulance staff milling around in the cold night, waiting for the off. At the stroke of midnight we all rush around hugging and exchanging greetings. Within minutes Amanda and I get the first call of the year.

'Oh no,' says Amanda, gesturing me over. 'Happy New Year, Lysa.' We stare at the screen as it flashes up just two words:

two horrible words you never want at any time, let alone for the first call of the year ahead.

'One under,' it says on the screen. That means someone's gone under a train.

'What kind of timing is this?' I ask as we drive off to the station. Luckily we're not far away and there's not much traffic around. We get out the rubber gloves – somehow we know we're going to need them. At the station a small knot of grim-faced staff and British Transport Police meet us. The story is, the train driver alerted them the minute he realised he'd hit something. It was a fast train and he was flying through, so he couldn't actually stop at the station. But it had happened literally seconds after the stroke of midnight. Spooky.

'We think he hit someone and they're still out there on the track,' we're informed.

It's a freezing-cold night with a full moon. So far the police haven't spotted anything. We all stand there on the silent, deserted platform, shining our torches on to the track, looking for the body. More police officers join us. But it's a waste of time. We can't really see anything from up here.

It takes a few minutes to agree a plan of operation with the police: Amanda and I will get down there on to the track, walk along and see if we can find the person. The power's already been switched off. There is, of course, a chance they'll still be alive. Given the speed of the train, it's a very slim chance – but one we still have to consider. There's no time to dwell on what sort of state the body might be in. Hastily we don our hi-viz yellow and green jackets with reflective strips.

And the hard hats. We keep all this stuff with us all the time, just in case we run into situations like this.

'This is the last time I'm working on New Year's Eve,' Amanda grumbles to me. A £20 note for every time I've said that and I could have retired to Italy by now. Though I never really mean it.

'Yeah, this is pretty grim,' I concur as we climb down. And so, as the rest of the world hugs, kisses, glugs down champagne and parties in the New Year on a cloud of optimism, we're on the track looking for… what? A mutilated body? A half-dead person? We haven't a clue. It's actually quite surreal. It feels a bit like we're in a movie – a horror flick. As we go along to the unlit section, the full moon shines right down on us.

'All we need is the werewolf,' I say to Amanda, trying to break up the bleakness of the situation. It's chill-your-bones cold.

'Yeah, this is just what you want to be doing tonight,' Amanda replies, as we strike out into the darkness.

'I'm not sure if I really want to see what I'm supposed to find.'

Then, about 100 yards along, torches trained firmly ahead of us, we stop. We've found something. It looks as if it may be a spleen. And incredibly it's steaming. It's still at body temperature but because tonight's such a cold night, the steam is rising up out of it. A little bit further along and there's more – this time it's a man's slipper. A bit further down and, oh, no! We're looking at an intestine – it's spilled out from the guy's stomach. This is a horror show. We don't talk now, we just keep going – and soon we find the rest of the man's body, strewn along the track.

What had actually happened was, he'd thrown himself on to the line and the impact of the train had virtually split his body in half. So when we finally get to what's left of him, the top and bottom half of his body are hanging on by skin only. And the contents of his abdomen spilled out along the track. Ironically the other slipper has remained firmly on the man's foot. He'd been standing there in his pyjamas. But the impact of the train has ripped them off.

You couldn't do this job if you were squeamish about things like this: you'd go mad. But seeing a human body ripped apart like that does make you feel very strange. The man had wound up face down on the track. So at least we didn't see his face. But what a brutally violent end. How desperate can a person be, in those last seconds, standing there, all alone in the freezing cold on a train platform, to end it all like this?

But now our job is over and we're ready for the next one. Of course we're shaken but there's no one needing our help and the police will organise the actual removal of the body parts from the track. Someone else has the terrible job of picking up every single bit of this man, sending it to the public mortuary and eventually the coroner. And someone else again will have the awful job of finding the man's family and telling them what's happened.

It was suicide, all right. He obviously meant it. When men do it, they tend to do it violently. Maybe he'd waited until just after midnight; maybe it was sheer coincidence. After all, how would someone know that the train was going to pass through just after midnight?

A week later and we find out that the man had severe mental health problems. He'd been in and out of psychiatric hospitals for years. He'd tried to kill himself before. But this time he was determined to get it right. And for me, the thought of that haunting walk in the stark, cold moonlight will hang around in my mind for a long, long time.

DEBT

The call is: male, 33, head injury. Suspended? That means he's in cardiac arrest. We tear through the streets, sirens screaming. At a terraced house we're met by a frantic young woman in her twenties. She's beside herself, distraught, barely coherent.

'He's upstairs, I think… he's… dead!' she wails at us as we dive up the stairs.

Two ambulances are normally dispatched to a 'suspended', but we're first to arrive and there is the man, lying on his back on the bathroom floor. Bathrooms are usually the place where most 999 fatalities occur, particularly when someone's had a heart attack. The first thing I notice is that the man has two black eyes. Blood is trickling from his nose and ear. This is all a bit odd.

'D'you know what happened?' I ask the girl. She's slim, with dark, curly hair. Naturally she's in a state – but she now seems incredibly nervous, holding her hand to her mouth like she doesn't want the words to come out.

'No, I just came home and found him like that. He must've been beaten up.'

We start to work on him, knowing it's probably useless, attempting to get him back to life with cardiac resus, pushing his chest, trying to get a needle into his arm to give him drugs and to pass a tube down his throat to get oxygen into his lungs.

'Doesn't look like he's been beaten up,' comments Dave, who's working with me. And he's right: everything around the man's body is in its place, nothing knocked off the shelves, no mess. And no blood anywhere other than the trickle. The girl stands there, silently willing us to end the nightmare. Does she think it might have been a break-in?

'Yeah, the front door was open when I came in. Must be. Will he be all right?'

No, love, he's never gonna be all right again, goes the voice in my head. The funny thing is, relatives or families always say this to you – even when it's glaringly obvious that the very opposite is true. Maybe they just want to reassure themselves – or perhaps they just don't know what else to say, human nature being the way it is.

But we go through the motions. 'He's pretty poorly at the moment,' I tell the girl, trying to let her know what she hasn't yet accepted. 'Essentially he's dead and we're trying to get his brain oxygenated until we get him to hospital – so if he recovers, there's less chance of brain damage. But we really don't know why he's like this, no idea at all.'

It's all very mysterious. What she's telling us and what we

see don't really add up. Dave goes back to the ambulance for some equipment. And when he comes back he confirms: there's no sign of disturbance anywhere in the house, which is easy to see because it's a pretty small place.

'It just doesn't look like someone's broken into this place, Lysa,' he says, shaking his head. Now the police, automatically alerted by the girl's 999 call, have turned up and are taking care of her. They'll take her with them to the hospital. But our job now is to get into the ambulance, speed up, race the man to hospital and carry out resus all the way in. Which is what we do. But by the time we get there it looks like it's all over. As we're clearing out the ambulance and doing our paperwork before we get the next job, one of the nurses comes out to us.

'Don't go,' she says. 'We've been talking to the police. That man's been shot.'

Shot? I'm flummoxed. The police will now need statements from us, what we saw when we got there, that sort of thing. But where was he shot? We would have seen it, wouldn't we? We didn't. What happened was the man shot himself in the temple and the wound was covered by his hair. And because the girl talked about him being beaten up, we didn't even look for a shotgun wound. We didn't suspect it – and so we didn't spot it. You hear a lot about gun crime nowadays but this is still pretty unusual. We don't get a lot of calls to people who have shot themselves. Yes, we go to a lot of armed incidents – which usually come to nothing. But this is strange. And there's no sign of a gun.

Later we get the full story from the police at the station where we go for statements and fingerprints. The man shot himself in the head with a small handgun. The swelling and bruising around his eyes were caused by the location of the bullet that killed him. But once they talked to her, the police weren't convinced by the girl's story, and when they grilled her further, the truth came tumbling out.

She came home and found him lying there in the bathroom with the gun, a lady's handgun, in his hand. And she panicked. She explained to police that she'd remembered that if you commit suicide you don't collect any life insurance. So she took the gun, hid it and when she rang 999 made up her story about him being beaten up by strangers.

Once she spilled the beans to the police they let her go. It was all about money. It turned out they'd had lots of debt. He'd obviously made up his mind that the gun was his only option. What made it worse for me was when the family turned up at the hospital before we went down to the police station, I recognised one of the men from my school days. I didn't talk to him – it was clearly a terrible time for all of them – but I remembered that as a teenager I'd known them both by sight.

So the man whose life we'd have tried to save, if we could, wasn't really a stranger. I just didn't recognise the corpse on the floor as one of the boys I'd discussed with my friends as a teenager. You know how you do this as a kid: even if you don't talk to certain boys, you eye them up endlessly as they do you, and you all have an opinion on the way they look or behave.

But things had gone very wrong for him, and he obviously couldn't see any other way out of the mess.

PURPLE PLUS

Paramedics don't as a rule expect to turn up to a house and find more than one person collapsed with cardiac arrest. It just doesn't happen very often. When we train for the job, this mantra is drummed into us: one patient, enter carefully. Two, be very careful indeed and consider carbon-monoxide poisoning or any other threat. Three people, do not enter at all, call for assistance. Or you could fall victim to the same fate yourself.

But here we are on an autumn day, called to a house where we already know that someone's gone into cardiac arrest – and control, say, a lone colleague, Rob, is already there. In situations like this it's normal for control to send in a second team because there can be a lot to do. Carrying out CPR can be hard work and, believe it or not, there are degrees of being dead following cardiac arrest.

Sometimes a person can be 'newly dead', which means there's a chance we can get them back – but we have to pull out all the stops and carry out CPR very quickly. Around 5–10 per cent of the time this can work so well that a person survives after a stint in hospital. But for the other 90–95 per cent of the 'newly dead', we do our best when we get there and carry out CPR while transferring the person to hospital. Then the hospital will continue trying for a little longer. If there is a blip of life, the person goes to intensive care, but in such cases the person usually dies, maybe a day or two later. However, there is one other term paramedics use which means no hope, very dead indeed: it's called 'purple plus'. In these cases, carrying out resus is a futile exercise – and doing so gives a false sense of hope to friends and family.

Today we've arrived to find the front door wide open. Inside we're immediately confronted by a woman lying on the floor of the living room, unconscious and barely breathing. But there's no sign of Rob. What's going on?

Then Rob appears, charging down the stairs. 'OK,' he says. 'I've got one purple plus upstairs – now this one's gone off on me.' The woman opened the door and directed Rob upstairs to her husband. Rob noted briefly that she looked a bit pale as he rushed to the man. Who, sadly, was very very dead. And stiff. Now Rob has to go down and tell the woman the bad news.

To his amazement, he found her unconscious, barely breathing – and a very unhealthy shade of mauve. Rob managed to lift her on to her back and insert an airway into her mouth – to help keep her tongue from the back of her throat and get air into her lungs – but he then had to retrieve some of his equipment he'd left upstairs. That's when we turned up.

So now three of us are trying to resuscitate the woman. But I'm wondering if maybe, just maybe, something has occurred in this house to cause two people in different parts of the house to go into cardiac arrest.

'It's just not normal for two people in the same place to collapse from cardiac arrest like this,' I say, as Rob and the other paramedic start CPR on the woman.

'Maybe it's carbon-monoxide poisoning,' suggests Rob. He's right. I go round the house and start to open up windows and doors, just in case. Then I go out to the ambulance and radio for another ambulance and the fire brigade. We can't afford to rule out carbon-monoxide poisoning. Because it has no taste, smell or colour, there is a chance that a faulty water heater or boiler has been emitting poisonous fumes into the house. Carbon-monoxide poisoning can cause cardiac arrest – and if someone already has a heart or respiratory problem, they're doubly at risk from it.

Back in the lounge, it's all hands on deck. Now the woman has gone into full cardiac arrest. She can't breathe for herself. Nor is her heart beating independently. We manage to get the tubes and lines into her to try to get oxygen into her lungs. Then I look up and see four figures, fully clad in protective gear and breathing through a noisy filter system. It's the fire brigade, kitted out in their Darth Vader gear, like something out of *Star Wars*.

'Get yourselves and the woman outside,' we're warned, so we get her on to a trolleybed and take her outside. Another ambulance crew turn up to help us. The fire brigade start to go through the house as we race off to the hospital, alerting them of our concerns about carbon-monoxide poisoning on the way. We manage to get a pulse back on the woman before she's handed over to the medical team. Then, after we've finished the job, we need to have blood tests, in case there's carbon monoxide in our systems – but luckily we're in the clear.

Later, sadly, we learn that she doesn't make it after being moved into intensive care. She was only 52. But the whole story remains a bit of a mystery to us. Afterwards we discover that some carbon monoxide was detected upstairs, around the man – but none was found around the woman downstairs. Apparently she'd been staying elsewhere, visiting a relative, until that morning. So she came home – and found her husband dead on their bedroom floor. He already had a heart condition and it's likely that the carbon monoxide finished him off. But as for his wife, it seems she discovered his body, managed to ring us – and then collapsed and died from shock. A truly gruesome coincidence. And a dreadfully tragic end to a marriage.

LOSS

The man has blood on his hands and all over his clothes. He stands at the door to the semi, a pleasant, middle-aged suburban man whose peaceful world has just been shattered into a million pieces.

'I think you're too late,' he says as we rush past him into the living room. 'He's gone.'

There is blood everywhere, on the sofa, on the floor, down the front of the man's clothes. He looks ghastly. John and I move the slumped body on to the floor. There's not a second to spare. We start to try to resuscitate him by compressing his chest to push blood around the body. We use a suction unit in an effort to clear blood that has accumulated in his throat. Using a three-piece item of equipment, a bag, valve and mask device, we put the mask over his face and squeeze the bag, hoping to get oxygen into his lungs. But it's hopeless – there's simply too much blood, and we can't stop it from occluding his airways. All we are doing is pushing it down into his lungs.

'We'd better change the bag, Lysa,' says John, and he's right, the whole thing is full of blood, so we put in a new one. It doesn't make much difference, he's just bleeding too fast. We can't get a tube into his airway either, there's just too much blood. We try, though we know it's pretty hopeless: we put his legs up on to the sofa cushions, to endeavour to return the blood in his body to his vital organs, maybe keep his blood pressure up a little. We get a cannula tube into a vein in his arm with drugs that might help his heart to restart. Then we start to get him out to the ambulance. We are going through the motions. The man is bleeding to death big time.

His name is Ian. The man in the bloodstained clothing is his father, Paul. He's been filling us in on the background while we work. Ian has been living with Paul, a widower, for about a year. Ian hadn't been too well for the last few days and wasn't eating much. But his dad thought he was OK.

'I just went upstairs to the toilet and came down and found him like this, on the sofa, with all the blood everywhere. For a minute I thought someone had come in and attacked him, there was so much blood. But then I realised the blood was coming from Ian's mouth.' Ian had been a heavy drinker since his teenage years. But whereas his friends started to work, settle down and cut back on the drinking sessions, Ian kept up the heavy drinking. And it continued through his adult life. His alcoholism had lost him friends, girlfriends, jobs – and eventually his wife and son. His house had been repossessed. Finally he'd had nowhere else to go but his dad's home.

'I knew he was in a bad way,' his father told us. 'The doctors had told him his drinking would kill him. But he couldn't stop.' The doctors weren't saying this lightly: they knew that Ian's liver was badly damaged by the alcohol. He was starting to look jaundiced and his skin and the whites of his eyes had taken on a yellow tinge. His stomach was distended and uncomfortable. He'd often have to leave his trousers undone at the top because they were too tight: all because his damaged liver had by now become enlarged and fluid was collecting in his stomach.

'What's happened to him?' Paul asks me, clearly in shock but still surprisingly composed, as we get into the ambulance.

'It looks like the veins in his oesophagus, the tube that runs from mouth to stomach, have ruptured. That's why there's been so much blood,' I tell him. 'It's very serious. We've done what we can but it's not looking good. He's lost too much blood.'

'You don't expect to bury your child, do you?' Paul says to me as we drive off. Unlike many relatives, he seems to have accepted the reality of Ian's situation, maybe because of the doctor's warnings, or maybe he's just a sensible sort of man.

'No, of course you don't. But they'll do whatever they can at the hospital.'

Technically Ian is dead. We're just keeping him ticking over. But even at the hospital they can do no more. Had the veins in Ian's oesophagus ruptured in a hospital setting, with surgeons and equipment at hand, they might have saved him. But away from a hospital, the results of such a rupture are devastating. Still wearing his bloodstained shirt and trousers, Paul waits for the final, worst bit of news in the relatives' anteroom while the doctors complete their paperwork.

It's a classic story of death from alcohol abuse. Because the damage to Ian's liver is so severe, his blood has lost its ability to clot properly. So he starts to bleed inside, slowly. The bleeding from the oesophagus has probably started a couple of days before, initially running down into Ian's stomach. He's probably ignored the dark, tarry-coloured stools that he was passing, the sign that the blood is running into the stomach. And so finally he has a catastrophic haemorrhage. In simpler terms, he has bled to death. At the age of forty-two, witnessed by his dad.

We're now on our way back to the ambulance station to clean up and change our uniforms. We're covered in Ian's blood. It's an unhappy night for us. But then I think of Paul going back to the neat suburban house, and the blood-spattered lounge awaiting him, the phone calls and explanations to shocked family and friends and the funeral of a man who probably never envisaged such a terrible, bloody end. Even when the doctors told him to stop.

KNIFE WOUND, YOUNG MALE

I'm in the car, waiting for the next call. But someone, somewhere has just been stabbed, a brutal act of violence that means their last moments of life are already ticking away.

Right now, a stranger's body is desperately using all its resources to try to stay alive. Their pulse is speeding up; fear and adrenalin have sent the pulse racing so that blood and oxygen can reach the brain and the other vital organs. And while this internal battle is being fought (and possibly already lost) a passer-by is frantically dialling 999, waiting for seconds that probably feel like minutes as the operator connects them to the ambulance service.

Then, no matter how much they might be panicking, they try to give a coherent idea of the exact location, tell the operator what they think has happened and, if they can, explain what condition the person lying there is in – a difficult thing to do at any time, let alone when every single second counts.

It's a rainy, cold night in Croydon. I'm the nearest ambulance response but precious minutes are already ticking away as the details of the job – knife wound, young male – are sent to me via the mobile data terminal in my car. Tonight it's busy and I'm miles away. I switch on the blue lights and the sirens and race through the rain, the traffic and the distracted pedestrians with a death wish who seem to keep stepping out in front of me with no sense or idea of the urgency I'm feeling inside.

Then, total frustration. I'm almost there but the control room comes through: they want me to wait around the corner from the address while they confirm that the police have already arrived; they can't risk me being attacked too. I'm already wearing my stab vest – a routine precaution – which would give me some protection in a worst-case scenario. In this business every risk, big or small, has to be taken very seriously. Ambulance crews have been knifed, stabbed, punched, kicked: most of us who've done it for years or worked in A&E have experienced our fair share of abuse from the public; it goes with the territory.

Waiting in the car, I start to prepare myself for what actions I'll need to take when I get there, but tonight, for some reason, I can't help imagining the scene that will await me. How bad will it be? How much blood have they lost? Will they still be conscious and, if so, what thoughts are going through their mind? Is he lying there, unable to move, thinking, I'll be OK, help'll be here soon, I can hang on? Or does he already sense that these are his last minutes of life, as the blood drains from his body? Who does this young man think of, wishing they were with them now to comfort and hold them instead of these strangers? His mum? His girlfriend? Is he thinking, So, this is how my life ends?

Then abruptly the call comes through confirming that the police are already there. I drive around the corner and, as is so often the case, I find a scene of unbelievable chaos out on the pavement. Screaming and distressed bystanders, who never in their lives dreamed they would actually witness such a shocking sight, helpless, tearful, afraid and even angry. All eyes turn to me as I arrive, carrying the equipment to help me look after the young man.

Seeing me, as happens so often, their hopes soar. The ambulance is here, they'll save him, they're thinking. I know from experience that unfortunately when knives are involved such hopes can be unrealistically high, and tonight this could be one job where there will be no good news for anyone. One young girl has been trying to comfort the boy, pressing her coat against the wound in a desperate effort to stem any bleeding. Another bystander, also a young woman, has started doing CPR. She tells me, through tears, that she did a first-aid course at work recently, but she never in a million years thought she'd have to use it and she's worried she's not doing it right.

'He was talking to me for a bit after it happened,' she sobs. 'But then he went quiet. He seemed to stop breathing just before you arrived.'

'You did a really good thing,' I tell her and tell her to keep going with the CPR while I open up my paramedic bag and take out my kit. I put on my gloves and go into my routine.

As the young man lies there, inert, wearing the fashionable clothes so carefully chosen this morning and now soaked through from the rain, I check for a pulse and breathing – any signs of life. There are none. So I continue basic life support. I compress the chest to squeeze the blood through the heart to the brain and the rest of the body. And as I do this I look down at the boy's face, now expressionless with eyes closed.

I carry on working, briefly thinking of the hopes, the dreams and the promise that this young, healthy person had, all snatched away in an instant, never to be fulfilled. Did he play sport, go clubbing, hang out with friends? Did he have younger brothers and sisters? Were there girlfriends? Who loved him, worried for him? And who did this boy love? The smell of the aftershave he's put on before going out for the night wafts up to me; briefly I'm disarmed recognising it as the same one my son wears.

Another ambulance crew arrive to help me. We don't talk much. All of us are silently hoping against hope for a bit of a miracle: if only we can get the boy breathing for himself again. We cut his clothes so we can put pads on his chest to monitor his heart rhythm, but the monitor is showing a flat line. I pass a tube into his throat to allow us to push oxygen into the lungs. This will then pass into his blood and travel around the body as we rhythmically compress the chest. I put a cannula into a large vein in his arm to provide a route for the powerful drugs that may, just, restart the heart. If we manage to achieve really good CPR and he survives there is less chance he'll be brain-damaged through lack of oxygen.

It's still pouring with rain as we lift the limp body up from the cold concrete on to our trolleybed and carefully move it to the ambulance. We connect up to all the monitors inside – but, despite our efforts, there are still no signs of life.

We alert the hospital that we are coming in and, throughout the journey as we rush, sirens blaring, through the rainswept streets, we continue to compress the chest, squeeze the bag that inflates the lungs with oxygen. At the hospital we take the trolley into the resuscitation room to the waiting medical team and help move the boy across to the hospital bed. We stick around for a while to help the staff with the resuscitation.

Tonight it's my job to book the young man in at the A&E reception. The receptionist hands me a card and I notice that the last time he was here in this hospital was when he was seven years old and had stomach pains. How different it all is this time.

It's a bad night. All our efforts to save the boy have been futile. The knife had plunged deep into the chest and right into the heart. He's gone. So after a while I go in to say goodbye to the body lying behind curtains in the resuscitation room. No longer dressed in fashionable clothes, he lies naked, covered by a hospital blanket.

Staff have removed all the tubes and needles; his body looks absolutely perfect lying there. There never was really much blood loss that we could see. All the damage and bleeding happened inside the chest. All you can see is a tiny wound on the left side of the chest, covered now by the blanket.

The police have gone to find the family, break the terrible news and bring them into the hospital to say, 'Goodbye.' Their lives will never be quite the same again after tonight's terrible news. I want to be gone before they arrive because, for me, these are the worst moments of all. What greater tragedy can a family endure than the sight that will greet them when they arrive here at the hospital?

I grab a coffee from the machine and go out to my car which the police drove back to the hospital, to clean and restock the equipment. I finish my paperwork. Then I talk to the police and answer their questions. I will be formally interviewed at a later time as they begin to build their case. Sadly this isn't the first time I've been called out to a fatal stabbing. No matter how distressing the circumstances – and this is one of the worst because there was so little we could do – our training is to do whatever we can when we get there. This time, however, by the time we'd arrived it was too late.

I'm back in work mode now, ready to take another call. Incredibly I'm called out to a house with a seven-year-old with tummy pains. It doesn't look like it's too serious. But as I sit there talking to the worried parents I look at the little child and I can't help feeling distracted by the thoughts of the other boy still floating around my mind. What will the future hold for this little one? On nights like these, when you've seen a young life cruelly snatched away so swiftly, you can't help but hope with all your heart that circumstances will be kinder to this child as they grow up in the world.

BREAKING THE NEWS

How I hate the 7pm–7am night shift, especially the first one in a row of four, like tonight. Even though I've done them, on and off, for years, I've never got used to the disruption to sleeping habits. Some people fall sleep easily when they get home. I can't. So tonight, even before the first call, I already feel tired. I tossed and turned last night and didn't get my eight hours. The prospect of 12 hours to go with God knows what happening out there isn't doing much for my mood.

The good thing is, tonight there's three of us in the ambulance: I'm with Tony, who's driving, and another guy I know and like, Jed, who's joining us on the shift to observe and learn. Jed's a trained nurse with a dry, Scottish wit. He's already starting to tell me a joke when our first call of the night comes through: 15-year-old girl, breathing difficulties.

'What do you think?' Jed asks, eager to read my mind, as Tony swings the ambulance out of the station.

'Hmm, could be asthma, could be a panic attack,' I say,

trying to reserve judgement until we get there while simultaneously thinking ahead. Calls do tend to fall into certain groups. But you try not to label until you've excluded all other possibilities.

We pull up at a pleasant terraced house and a woman in her thirties opens the door. 'She seems pretty calm, maybe it is a panic attack,' I note to Tony.

But there's nothing calm about the 15-year-old, Susan, sitting on the stairs, crying, breathing very, very fast and wheezing. It's not good. She's clearly terrified by what is happening to her. Clearly something is affecting her breathing. Tony's getting some oxygen on her so I can take a look at her and figure out the rhythm of her breathing. Is it regular? Is it laboured? Her skin colour is quite pink – a good sign – but the wheezing sound when she breathes out isn't right in anyone's book. Stethoscope out, I listen to her chest to see if the air is getting into all parts of her lungs. Maybe part of her lung has collapsed? Maybe not, as I can hear air going through, but she's having to struggle to breathe out.

Jed's asking Mum, 'Has this happened before?'

'She's asthmatic and takes Ventolin but she's never been this bad before,' she says calmly. We decide to give her Ventolin from our kit to help the wheezing – a fine mist which goes from the oxygen mask into her mouth and nose to try to help relax the airways. But it doesn't help much. And now she's getting really agitated.

'I can't breathe, I can't breathe,' she gasps, which must be horribly scary with a mask on your face. Jed's already gone to get the carrychair. Somehow, panicky as she is, we manage to get her in it, blanket around her, mask still on.

But things are getting worse by the minute. We can't calm her down. She's very distressed and is now having a real problem breathing at all – a sure sign that not enough oxygen is getting to her brain. Mum scuttles round, getting her keys and jacket, still surprisingly calm as we get Susan into the ambulance and try to get her to sit on the trolleybed. But she won't sit back on the bed. She's kneeling up on it, facing the back door, distressed beyond belief.

'I can't breathe, I'm going to die, I'm going to die,' she wheezes. This is awful.

I do my best to help her sitting on the foot of the trolleybed.

'Look at me, Susan, you've got to save your breath – we're not going to let you die,' I tell her. It's the only thing you can say in such circumstances. After all, that's what we're here for. But my words are not helping. As we race off into the night, she's half-sobbing, half-shouting between desperate, rasping, wheezy breaths. 'I can't breathe! Don't let me die!'

Pulling the mask off, she's now totally hysterical with fear. Then without warning, she smacks me round the face, full on. 'I'm going to die!' she yells. She's tiny and small-boned in her jeans, trainers and t-shirt. But I still feel the effect of the blow.

'Look, sweetheart, you need to keep a grip on this. It's worse if you keep shouting,' I reason. But nothing is helping. She's frantic with fear. Her mother doesn't say much, just sits there looking helplessly at her daughter. She probably never imagined it would get as bad as this. I ask Jed to put in a priority call to the nearest hospital and to tell them it's developed into severe breathing difficulties. We'll be there in a matter of minutes. Now we're in a 100 per cent emergency situation. And there'll be a doctor waiting for us when we get there.

By now it's a struggle to keep the probe on, the mask in place and my balance in the vehicle. Suddenly I remember a similar call I went to last year: a 12-year-old asthmatic boy in similar circumstances who had died. He too was shouting, struggling to breathe, and you could hear him down the other end of the hospital corridor. It haunted me for weeks afterwards.

'Please don't let this one end up the same way,' I pray silently, knowing all too well that asthmatic children or youngsters are always a real worry because they have the potential to deteriorate so quickly. But we're nearly there. Jed is shouting out warnings now to Tony, who's driving, 'Speed bumps ahead', 'Two minutes away,' 'One minute away.' His directions give me a chance to get some equipment off and make a rapid exit. Through the journey, Jed and I manage to catch each other's eyes a couple of times. We do not need words to acknowledge that this is a worst-case scenario.

We've arrived. Jed rushes out to open the back doors and, yes, the team are waiting for us. She's still gasping, 'Don't let me die!' and the waiting doctor gives me a knowing glance which I take to mean that he's considering the possibility of a panic attack. Have I got it wrong? I hope like hell I have: it'd be reassuring for everyone. We follow him down the corridor. The girl is still kneeling up on the trolleybed in an attempt to breathe, still frenetic with fear, despite everyone's efforts to calm her down.

Mum still doesn't say much. Occasionally she tries something like, 'They're doing their best, love, try to calm down,' but she seems to have gone into slow motion. How unreal it must seem to her. A few of hours ago her priority was cooking the evening dinner. Now her daughter's life hangs by a slender thread. We're in the hospital resus room. We tend to let parents stay with kids – it can be hard not to – and we help the team try to connect various monitors to the girl. As I try to put a blood pressure cuff on her, she manages to whack me in the stomach, still resisting our efforts.

Suddenly, oh no, she's lost consciousness, her arms drop, her eyes roll back in her head, she falls back – and pees herself, soaking her jeans right through. Her body has tried its best to compensate for the lack of oxygen but all her resources are used up. There's just not enough oxygen in her brain to keep her conscious. So her body has simply let go. When that happens, it means you're truly out for the count. All your bodily functions relax to the point where it happens.

'Oh, shit, respiratory arrest,' someone says.

Thank God we got here, I think. It's looking dire. But at least the right resources are all here in the hospital. As the hospital's cardiac arrest call goes out, we're desperately trying to get oxygen into the girl with a bag and mask. The doctor passes a tube into her throat, trying to secure her airways and protect it from vomit. The stomach muscles too are completely relaxed, so vomit too will often come out in this situation – one more occupational hazard for emergency teams.

Now the cardiac arrest team, paediatric staff and an anaesthetist, seven people in all, run into resus to start to work on her. You need two things to stay alive: a pulse and the ability to still breathe. Her breathing's stopped but she's still got a pulse, so there's the faintest chance. Her mum stands there, her hand over her mouth, just staring but saying nothing. Relatives don't always appreciate how bad things are. Why should they? They've never been in this situation before, so they've got nothing to measure it against.

At times like this some families will insist on staying, watching everything that's going on, no matter what. But, partly because I sense that she'll agree, partly because I desperately want to spare her the worst, I offer to take the girl's mum into a side room. And she agrees. Apart from writing up the paperwork with Jed, my role in this drama is over, whatever the outcome. So we sit in the relatives' waiting room and I ask her about Susan, trying to distract her. To talk about anything else but her daughter right now wouldn't work.

I ask the normal questions for a time like this, those that every parent responds to: 'What's she like at school?' 'Does she have a lot of friends?'

Yes, she loves netball, she's got many friends. 'She's a good girl, Susan, no trouble at all,' she tells me sadly. I manage to fetch her a cup of tea from a machine outside. She's just about to drink it when a doctor and nurse walk in.

I know instantly what's coming next. And so does Susan's mum. She looks at them and shakes her head. 'No,' she says. 'Don't talk to me.'

'I don't want to hear what you have to say,' she pleads, her face in her hands. She's touching her face as if to assure herself this is real, that she's not having a bad dream. Then she turns to the wall. Her voice is still quiet but it's breaking. 'I can't hear what you've got to say. Please don't tell me.'

My arm goes round her. I'm a total stranger and yet here we are together in this terrible situation, a moment in time that splits her world into pieces. 'Look, Carole, you have to listen,' I tell her. Then she starts sobbing her heart out. And I'm close to tears too. We might witness this kind of thing a thousand times over as part of the job – but that doesn't stop the tears welling up or harden you to someone else's tragedy. Then the doctor goes into his practised speech. 'I need to tell you that despite everything we've tried your daughter hasn't responded,' he says.

In fact the team are still doing chest compressions as he talks to us. They know it's all over but they're buying time so the doctor can tell the mum. And they want to give her the chance to come in and hold her daughter's hand before they cease completely.

So we go back in together. Everyone stands back to leave us with Susan. There's still a tube in her mouth but she looks peaceful, normal even. And the mother is hugging her child, clinging to her, sobbing her heart out. At that point I have to walk away. I can't witness this any more and I can't stop my tears. I go to the loo and splash my face, somehow managing to compose myself. Then I go back outside to the ambulance. Tony and Jed have already sorted everything out, checked and replaced equipment.

Normally after a job like that you get a call out to a really ordinary job, a twisted ankle, or something fairly innocuous. When you turn up you're ready to shout at the person, 'You think you've got a problem, well, you should have seen what I've just seen,' but of course you say nothing. And anyway you still don't know what the job after this one will bring.

RELATIONSHIPS

A CRISIS

A 22-year-old man has overdosed. I turn up to be greeted by a young woman holding a baby. 'It's my boyfriend,' she informs me. 'And he's pissed off that I called you.' Then she ducks into another room. And shuts the door.

I go into a tiny bedroom, almost taken up entirely by a bed. The young man in it looks at me. She's telling the truth. He certainly doesn't look as if he wants anyone to help him.

'For fuck's sake, what're you doing here?' he snarls, then turns on to his side, dragging the covers over his head.

'I'm Lysa. Your girlfriend's worried about you. I've just come to check you're OK,' I say cheerfully. 'What's happened?'

'Took an overdose, didn't I?' He's angry. There's no contrition in his voice.

'What've you taken?'

'Them,' he says, gesturing to a small packet of ibuprofen by the bed.

'How many?'

'All of 'em.'

The packet is empty. But there were just eight pills inside. So this doesn't look too bad. The chances are it won't be too harmful, which is just as well as I'm working on my own. How is he feeling now?

'Shit, aren't I? Why're you still here?' Hmm. He's snappy, not in the least interested in being helped in any way. It's been a busy day. Part of me feels like leaving him to it, walking away. He's got bad attitude – and there are a lot of people out there needing our help.

But I have to try. 'Look, if you don't want me here, I'll go. But could you please let me take a few details, just to make sure that what you've taken isn't going to be harmful?'

Grudgingly he agrees. He gives me a bit of background information, his general state of health, medication he normally takes. One thing I am obliged to do in a situation like this is ring through to a central poisons unit, just to confirm that the overdose wasn't too serious or harmful. I know it wasn't but the formalities still have to be observed. So I start to ring the number. But now he's interrupting me as I do so.

'I wanna word with you,' he gestures. Strangely he's a bit calmer now, less upset. Perhaps he's stopped feeling edgy because it's obvious he hasn't done something serious. The call can wait.

'What did you want to tell me?' I ask.

His response is to start crying. Just like that, from tetchy to tearful in a few seconds. I stare at him, puzzled. He's really distressed: something is hurting bad.

'I'm sorry,' he sobs. 'I didn't mean to have a go at you. I just feel terrible. And I didn't know what to do.' Then the whole story comes tumbling out. And it's a very sad state of affairs.

Two months before, his girlfriend, the young mum, had been seriously assaulted. In a dark street a small group of yobs had attacked her, each one taking turns to rape her. The case hadn't even been to court yet. And the gang had been charged. But for some reason he seemed to think they were going to get off. And he felt both devastated and helpless about what had happened to her.

'I let her down,' he tells me. 'I just know she wants to leave me and take the baby because she thinks I've let her down. And if they go, I don't know what I'll do, I can't bear the thought of living without them,' he says and starts crying all over again, as if voicing his worst fears will somehow make it happen.

You have to feel sorry for this young couple. Rape, in itself, is devastating in any circumstances. But the fact that this woman's partner has even thought of harming himself, albeit half-heartedly, makes it even worse. He hasn't been coping at work, either. A delivery man, he's just lost his job because he wasn't getting to work on time. And he's started drinking too much, as a result getting into debt. In telling me his story he's gone from being a vicious, nasty sod to a fragile, vulnerable young man who believes he's going to lose everything – through no fault of his own. He's told no one about all this. He's kept it all bottled up. And he can't bring himself to discuss it openly with the girl.

'I love her so much, I don't know what I'm gonna do if she goes,' he keeps saying.

'Has she said she wants to go?'

'No. But she will. I know she will.'

In the time it's taken for him to pour his heart out to me, I've sensed that the girl is hovering outside the door. Is it OK for me to ask his girlfriend in, so we can all have a chat? He sits up and blows his nose, and I usher her in. She's still holding the baby.

'Is it OK for me to tell her what you've told me?' I say. He nods.

So I tell her everything, what he's been feeling, why he's so upset. And that he's managing to blame himself for what happened to her.

'Oh, Phil, I don't blame you,' the girl says, perched on their bed, cuddling the baby.

'He's afraid all this will make you want to leave him,' I add gently, knowing full well by now that the guy's fears are groundless – he just couldn't voice them to anyone until now.

'I love you, you idiot,' she says, her own tears welling up. 'I love you, we've got a son, why would I leave you? We can get over this,' she adds sensibly.

Now they're talking properly. He's worried about the loss of his job.

She's assuring him that this setback won't affect them, won't send her heading for the door. 'And anyway you've been offered another job.'

Mentally he probably hadn't been confident enough to

accept the job. But now, at least, everything's out in the open and they can start to plan ahead.

I take the baby from her and leave the room, cuddling it, a cute little boy, about six months old, chubby with big blue eyes. Now they're talking, comforting each other. The pressure's off. Five minutes later I go back in and he looks normal, the tension's left his face. They're sitting on the bed, holding hands.

The poisons unit confirm that the overdose wasn't serious enough for him to need to go to hospital. But the pair give me their GP details so they can be referred on for counselling, if they wish. Their crisis, if you can call it that, had simmered under the surface all those weeks in that tiny one-bedroom flat. They'd lived together, not talking properly, she mistakenly believing he was pushing her away because of the rape, that he couldn't stand to go near her any more. And he'd pushed her away, blocked her out from fear that she'd leave – and because somehow he'd taken part of the blame for the terrible incident. Like so many young men, he didn't attempt to confide in anyone – he was too ashamed to talk about his feelings. But coming into contact with a total stranger, sympathetic and female, brought the barriers down. But when you think about what had happened to his girlfriend, it's virtually the worst scenario a woman can imagine. Is it so surprising that they were struggling to cope with it? To this day I'm convinced they made it, survived the court case and got on with their life as a loving family. They just needed to break down that terrible wall of silence that had sprung up between them.

Our job isn't always medical. Occasionally it's about communication and the lack of it. Sometimes people are lonely and they just need to talk for a few minutes. We're not here to be counsellors or mediators: our job is to help in an emergency. But there are times when, in order to help, all you can do is listen. So I'm glad I stuck with it.

LOST

Tonight my husband Steve and I are out working together. We rarely do this. But it's New Year's Eve, the children are staying with a relative and Steve already had this shift booked when someone else dropped out. So I've volunteered. At least we get to see midnight in together – whatever mayhem we have to cope with afterwards.

So far it's been a fairly normal start to the New Year: a call to a young girl in Croydon, lying in a shopping mall, rolling around in her own vomit, skirt round her hips. She's managed to pee herself too. Her friends are insisting, as usually happens, that she'd 'only had a couple of drinks'. Well, at least it's not that hairy old chestnut 'her drink's been spiked'. It's amazing how many times we hear that one. I know it happens. But not at the incredible rate it seems to strike people down in these parts.

Once we've got her into A&E – already resembling something close to a scene from hell – it's straight to the next call. But this isn't a typical New Year call: 24-year-old girl bleeding PV. That means she's bleeding from the vagina. In a nightclub.

The location, of course, is what you'd expect tonight. It could be serious – but then it might be nothing. You wouldn't believe how many jobs I've been on relating to a bleeding PV where you wind up asking, 'When was your last period?' only to be told, 'Oh, about four weeks ago.' Maybe it's painful or heavier than usual but essentially they've called 999 because they're having a normal period! Some women seem to think that if they call an ambulance they're entitled to a free pregnancy test. Please don't ask me why. But in this case we're told that the girl is ten weeks pregnant. So it may mean miscarriage – and an upsetting scenario. Rotten timing too – though, when you think about it, there's never a right time for losing a baby.

At the door, staff usher us through the heaving club. It's noisy, smoky and jam-packed: most of the area's under-thirties seem to be here tonight, pissed and partying as if their lives depend on it. The racket is deafening and it's a bunfight to get through to the girl. The golden rule of paramedic work is the person you're rushing to treat is always at the farthest end of the building. Or on the top floor. But when we get to the ladies' loo we face a very sad sight. The girl is sitting on the floor inside a cubicle. She's tearful and very pale. There's quite a lot of blood on the floor – and loads of blood-stained loo roll lying everywhere.

Steve stands outside. In a situation such as a miscarriage, if a male–female emergency team turn up, the woman will always take over. I take the girl's pulse and blood pressure. She's got a wad of blood-soaked tissue between her legs.

'I think the baby's come out,' she says, but all I can see are large clots of blood. She's probably about to miscarry.

She cries and cries. 'My baby… I can't believe it.' Words that strike at the heart, no matter how many times you may have heard them. I'm doing what I can to comfort her. Outside, Steve has been reassuring the girl's boyfriend, who is quietly waiting, sobering up fast. And two young girlfriends are here in the loo, trying to soothe her, be helpful – as much as you can help another woman at a time like this.

I go outside. 'Can you get me a chair, Steve?' I say to my husband, who now seems distracted by the passing parade of short, tight skirts in various stages of merriment. He weaves his way through the crowd to the ambulance and a few minutes later we've got the girl in the chair. We can go. But then, in all the blood and mess, I spot the foetus: a tiny, bloody sac, a little wet bag. Just an hour ago this was the beginning of a new life. The sac is very small, the foetus only is approximately 3cm long. I carefully wrap it in a small plastic bag as it will need to be sent to the pathology laboratory.

Seeing it, I can't help recalling other, similar situations. Like the time a girl just calmly handed me the foetus, wrapped in a toilet roll, when I arrived. She was very matter-of-fact about it all. Not like this girl. For parents-to-be, these moments are the worst, realising you'll never get to see your baby's face – or watch it smile at you for the very first time. Death before life even begins, a small tragedy in the general scheme of things – but a gaping loss nonetheless.

Now we're taking her outside. The contrast between this tiny tragedy and the scene that greets us couldn't be more poignant. Around us the whole world seems to be celebrating noisily, welcoming the beginning of a new year. But for this girl, her smart, black party dress rumpled and bloodstained and her boyfriend pale and shaken by the drama, the happy partying marks a sad ending. What an awful shock it is for them, youngsters in their twenties with everything to look forward to.

At the ambulance, her friends hug and kiss her goodbye. She's on the trolleybed, her boyfriend's seat belted into the chair next to her. They're holding hands. She's still crying a little. Steve and I check her blood pressure and pulse again. Physically she's fine; she's just devastated by the speed of events. We drive off to the hospital.

Soon Steve and I are rushing to the next call: more drunken mayhem, a fight between two brothers, broken glass, weeping girlfriends and more chaos in A&E. But things are starting to quieten down. Already the streets are emptying fast. It's raining and in a few hours the dawn of the new year will be coming up. Who knows? Maybe the girl will go on to have two or three kids with her boyfriend. Maybe not. But whatever the future holds, I'm sure she will never quite forget this particular New Year's Eve. The night she lost her baby.

A SAD MISTAKE

The house is on the road where I grew up, just a few doors down from our old house. I'm wondering if it's someone I might recognise from years ago – a friend of my mum perhaps? – as I drive to the call. It's a collapsed-behind-locked-doors call, a man, 82.

The police have beaten me to it this evening. It's quite a large house, semi-detached and on a corner. There's a big extension built on the side. I don't recognise it – but then it's a long time since we lived there. The door's wide open and the place seems to be swarming with police. Two are upstairs, moving around from room to room, and another two are downstairs, calling out as they walk around. Presumably they're looking for the man. It's odd that they can't locate him.

Then I hear a female voice crying, 'Please! Please! Will someone please listen to me!' The voice is coming from a back room on the ground floor. I go in. I find an elderly woman lying in bed. She's quite agitated and you can see she's upset, even though she looks perfectly well. 'Why won't anyone listen to me?' she bleats.

'I'm sorry about all this,' I say. 'But we're trying to find the man as quickly as we can, because we think it may be an emergency. No one meant to ignore you. Do you live here with your husband?'

'No! That's what I'm trying to tell them – but no one will listen! My husband is dead!' Poor woman. Maybe she witnessed the man wandering around, showing signs of distress – and, though she can't see him, she's assumed the very worst. What a shock for her.

'I do understand. But it's really important that we get to him as quickly as possible to see if we can help him.'

'No, you don't understand! He's been dead for years!'

Upstairs I can hear the police thundering around, opening and closing doors. And an ambulance crew and a paramedic in a car, Paul, are also here. The house is being overrun by emergency teams. But I suspect the old lady is telling the truth.

'Can you tell me what you mean about your husband, please?' I ask her.

'Look,' she says. 'There's been a terrible mistake. It's very embarrassing. Please tell them all to go.' I hesitate. But no one's found a collapsed man.

'Give me a minute and I'll come back and talk to you,' I tell her. Then I go the police and explain that the search is probably off, though they say they'll wait out in the garden while I talk to her. The ambulance crew race off – they've got a more urgent call. But I enlist Paul to come with me to find out the truth from the old lady. We sit down in the bedroom and she starts to tell us her story. Her name is Jean. She's genuinely sorry about all the fuss, all these people rushing into her home like this.

'But my husband died two years ago and ever since I've lived here on my own.'

Jean's children emigrated to Australia many years before. Now the house, once busy with the comings and goings of a lively, growing family, stands silent and still most of the time. You don't need to look too much around the place to realise that it is beautifully clean and tidy, a loved and cared-for home. Due to her chronic rheumatoid arthritis, however, it isn't Jean who keeps it spick and span any more. She's now virtually housebound. But an army of carers and cleaners come in at regular times throughout the day to help with meals, housework, cleaning. Apart from her children's annual visit, there are few other visitors.

'I've outlived most of my friends,' she tells us sadly.

Considering the fact that she's living all alone in this big family house, Jean's got everything well organised. The phone and other useful items are kept close to the bed. The light switches are even on timers. But, of course, she's desperately lonely. She hasn't got over the loss of her husband, Ron. They'd enjoyed a long and happy marriage for over fifty years.

'Ron and I never had a cross word in all those years. We lived for each other and our children,' she says. Sadly Ron died suddenly from an unexpected heart attack. And, because

Jean has so much time on her hands now, she spends a lot of time daydreaming, remembering better times. And sometimes, Jean explains to us, she misses Ron so much that she imagines seeing him right there in the room with her.

'I know that sounds crazy. But when it happens it seems so real, I feel so comforted knowing he's there. It doesn't upset me. It makes me feel happy.'

'So what happened today, Jean?'

'Well, I thought I could see Ron – but he was lying on the floor, next to the bed. So I thought I'd better call for an ambulance.' Once she'd got through to the operator, Jean had been asked if Ron was conscious and breathing.

'I turned to look at Ron – and of course, he was no longer there. What a silly fool! But how could I tell the woman the truth: that I only imagined my husband lying on the floor? The woman would think I was completely mad. So I hung up.'

What Jean didn't realise was the call had been processed: the emergency response had swung into action – and help was already on the way. I pop out to the garden and tell police the sad little story. Then Paul and I make Jean a cup of tea. We need to double-check that she isn't suffering from acute confusion or any other medical condition that might need treatment. Sometimes a urinary-tract or chest infection can cause an elderly person to become confused, though they're easily treated with antibiotics. But after we've checked Jean out and taken her pulse and blood pressure it's obvious she's perfectly well. The trouble is, like so many people, she's been trying to soldier on without any outside support for her bereavement, such as counselling.

Jean might be old – but she's got a lot of insight. She knows perfectly well that her husband is dead, but the hours spent daydreaming about the happy years of the past have become dangerously real to her. She just hasn't been able to come to terms with her loss. I ask Jean if she'd be interested in having some bereavement counselling.

'If you think it helps, dear, I'll do it,' she tells us. I promise to try to organise it through her GP. It is, I reflect, as we leave the big, silent house, a sad situation when memory can overwhelm reality to bring some sort of peace of mind or comfort.

'The lines just got crossed, didn't they?' I say to Paul.

Paul's not as sympathetic. He's on his third marriage. He shakes his head. 'Well… she's definitely not gaga. But I dunno. Fifty years, it's a long time. Can't see any of my exes imagining I'm still there, Lys, when I'm gone. It'll be more a case of, "Thank God the bugger's finally kicked the bucket."'

NO DOCTORS, NO HOSPITAL

The bedroom door is locked. I knock again. 'Ambulance service here, can we come in, please?'

Silence. Then, 'It's OK. I don't need you, you can go away. He shouldn't have called you!'

Christine and I exchange glances. We've been called out to an elderly lady with shortness of breath. The woman's son, a muscular twenty-something in a tight t-shirt, a real gym bunny, is surly and monosyllabic when he lets us in.

'Upstairs,' is all he says to us, though he's the one that made the call. Usually the caller wants to bombard us with information. But now we're trying to barter our way through the door. Christine tries next. She explains to the woman, Eileen, that we're happy to leave.

'But we just need to see you to make sure you're OK.'

Nothing, not a word. The door stays locked. Then, downstairs, a phone rings. We hear shouting, swearing, angry conversation. Then the son demands that we come down.

'Talk to my brother, will ya?'

Christine goes down, while I continue to try to persuade Eileen to open the door. She still doesn't respond. I'm getting a bit worried: has she collapsed since we got here? Is she lying there unconscious, needing resus? We just don't know.

'We've got a problem here,' says Christine, coming up on the landing. 'The other son sounded furious. He insisted we can't take his mum out of the house until he gets here.' He's informed Christine, in no uncertain terms, that he knows his mum wouldn't want to go to hospital under any circumstances. Not only are we strictly forbidden to take her, but if we try to take her away before he turns up, he threatens, there'll be trouble, big trouble. He says he's only five minutes away.

Right now his threats mean nothing. We can't do a thing. 'Lysa, there's something really bad going on here,' says Christine. 'He was really scary, it was horrible.'

We stand there perplexed. We can't leave the house until we at least see the woman. Is it a time-waster call? Or does this woman really need urgent emergency treatment? We're working in the dark.

Then a surprise. The door opens. Eileen looks shockingly ill. She's as white as a ghost and her hair is all matted. You've only got to look at her to know there's a major problem with her health. She falls back into bed, as if the very effort of opening the door for us has exhausted her. Her lips and tongue are a pale, creamy colour, indicating serious anaemia: lack of red blood cells carrying oxygen around the body. And she's worryingly short of breath. Quickly we get to work. First a blood-sugar test, a finger prick. But when I go to squeeze the blood from Eileen's fingertips, it's not red blood that's coming out, it's like a clear, straw-coloured fluid. It's the first time in my working life I've ever seen this. Christine is shocked too. She hasn't seen this before either. What the hell is going on? I don't know it then but later I discover that Eileen has virtually no red blood cells at all left in her body because of her illness. And she's been bleeding internally. A lot. Then, with a finger probe, I test her blood-oxygen levels: unreadable. She needs hospital, doctors, tests, treatment… *now*.

'Let's give her oxygen,' I tell Christine. But no sooner have we got the oxygen mask on to Eileen than we can hear the noise of the other son arriving downstairs. And there's a loud and furious argument going on between the brothers. It sounds like one of them wants their mum to go with us. The other one is adamant: she's not going anywhere.

'There's no fucking way Mum is going in that ambulance!'

'But she has to go, Dave, she needs to be in hospital!'

Eileen can obviously hear the arguing too. I explain that we don't exactly know what's wrong with her but the hospital will be able to help her. But now, as sick as she is, she's digging her heels in, between short gasps.

'I'll be fine here, dear,' she rasps.

'Why don't you want to go?'

'Well, dear, my baby' – she gasps – 'died in hospital. Years ago. I don't trust hospitals.'

She then informs us she's never been anywhere near a

hospital or a doctor since that time. Now things are starting to fall into place. The woman hates hospitals, blames them for the death of her baby, a traumatic time she's never recovered from. But now one of her children can see that whatever's gone wrong with her it's reached a really serious stage. In any event, the arguing is getting louder – and angrier. Now the two brothers are up here outside on the landing, going at it hammer and tongs. There's a sound of scuffling: it's getting violent.

We may not be strangers to this kind of thing but Christine and I are still human. And female. A paramedic uniform doesn't leave you immune to feeling threatened. We're here with two young, fit men, getting increasingly nasty. They could, if they chose, easily overpower us. Eileen too is getting more distressed. Christine goes to the bedroom door, hoping to calm them down.

At that point one of them explodes. 'If anyone takes my mum out of this house, this is what I'll do to them!' he screams, punching a hole in the opposite bedroom door. Then he turns on the brother again and they're pushing and shoving each other, yelling every kind of abuse in the book. It's frightening. And painful to watch.

'She'll die if she goes to that bloody place!'

'She will die if you stop her!'

Then the older, really stroppy one stomps into the bedroom.

'You can do whatever my mum needs here, can't you? You can't take her!'

Behind him, the other one is still trying to make him see reason.

'Let 'em do their bloody job, Dave!'

What a nightmare. The woman needs to be in hospital right now. But our job doesn't permit us to take anyone away against their will – let alone in the face of two sparring siblings on the edge of violence.

'Look,' I say as calmly as I can. 'We're not doctors. Your mum urgently needs a blood transfusion and we can't do that here.' For a brief second I think, Yes! I've got through – until the stroppy one throws another mad punch at the wardrobe. Whack! Another hole. This can't go on. We're being intimidated, Eileen is in a very bad way and these two lunatics can't see sense. Now the poor woman is wasting precious breath, trying to reason with them.

'Don't be angry, they're only doing their job.'

Then Christine looks at me. 'Lysa, do you want me to get out that bag from the vehicle?' she says.

She's affecting a normal, 'Shall I make you a cup of tea, dear?' tone. But I know she's as scared of these two as I am. Thanks, Christine. That's our signal. We agreed it ages ago. If either of us finds ourselves in someone's home in a dodgy situation just like this, 'Get out that bag' means 'Get out of the house *now* and get on the radio for the police.'

The brothers are standing between us and the bedroom door, the only route for an easy exit. The angry one just glowers, but he lets Christine through. As she radios outside for the police, the two are still yelling at each other, still incensed by the situation. It's both frightening and sad to witness because it's obvious this woman should have seen a doctor ages ago. They love their mother. Maybe they grew up with the story of the brother they never knew. So they've kept to her wishes. But now they're just terribly scared for her, though one has a bit more sense than the other. It's as simple as that. But even more painful is the sound of this woman, desperately ill, barely able to breathe, pleading with her boys to stop fighting. It's awful.

Within a few minutes there are police sirens outside. Calls like this usually mean the police converge all units to get someone there quickly. The boys go charging downstairs to be greeted by the police running up the driveway. Then police and sons clash. The brothers try to stop the police from coming in the house but they're outnumbered. Within seconds the police have got in, tackled them – and we can get on with the job and move Eileen out to hospital.

She's still protesting as we start to ready her, but the protests are feeble now. 'I'll die if you take me there,' she says in a half-whisper.

'Sweetheart, you'll die if you don't go,' I say. I never like talking like this. But the fracas has wasted precious time. And she's now exhausted by it all.

'Then you'd better take me,' she says sadly. On to the chair and downstairs to the ambulance and Christine. Somehow the police have calmed the boys down. We're ready to drive off when one of the coppers pokes his head round the ambulance door.

'OK if the boys go with you?' he says.

'No thanks,' I retort. We've been terrorised by these guys, 20 minutes of hell we could have lived without. No, they haven't attacked us – but they were that close. So the police take them and follow us to the hospital, where the woman goes into resus. As we come out, relieved and ready to start up for the next job, we run into the two sons outside, having a ciggy. They stop us. Amazingly they can't understand why we wouldn't let them come in the ambulance.

'You were getting so angry, we thought you'd hit us,' ventures Christine. I'm speechless at their insensitivity.

'I'd never hit a woman,' replies the really stroppy one. OK, give him the benefit of the doubt, though you can never be entirely sure. And later the full story reaches us. The woman had a massive cancerous tumour in her bowel. She'd been bleeding into her stools, possibly for years. Yet because it was such a gradual deterioration, somehow her body had accommodated it. The hospital treated her, gave her the necessary blood transfusions – and she was allowed to go home. I'm sure that whatever happened to her after that, her sons were there for her. They had to accept it. The trouble is, they were scared witless. When people get to a point where they can't cope with a situation, they might call for help. But sometimes they don't even like the idea they've actually asked for help. And if she'd been telling them for years, 'No doctors, no hospitals,' is it so surprising that they'd wanted to respect that wish? It was a battle of conscience between the boys, if you like. And, of course, young men can get very angry sometimes.

BEST FRIENDS

It's a very ugly image. The girl is horribly discoloured. There's a purply tinge to her face where the blood has been squeezed by the belt she's used as a ligature. She's wearing just her nightie. Her eyes are bulging, her tongue protruding, her legs look bruised and mottled but in fact it's post-mortem staining. And rigor mortis has already set in: she's cold and stiff, dead for quite a while. I call control to say the ambulance will not be needed. The police are on their way and they will remove the body for the post-mortem because it's what's termed 'a sudden and unexpected death'.

Being called out to a suicide is always distressing; when it's a young person it's bad news for everyone, especially families. Usually it's difficult to comprehend, especially when there's no obvious reason. For instance, why would an intelligent teenager in a comfortable home with caring parents and a bright future do this? No one has the answer.

People blame websites and chatrooms glamorising death where kids feel that they can somehow get their five minutes of fame if they join this macabre club. But I'm not sure. Sometimes drugs or alcohol might be involved. Or mental illness might be behind it. But not every time, not by a long shot. And there are times when you get a youngster who just can't see any way ahead because of the past, a past so terrible and haunting that ending it all, no matter how violently, seems like the only way out.

It's between Christmas and New Year, the time when London empties and you can actually drive around without being hassled by endless traffic. Alone in the car, reminding myself to check next week's shift, I'm jolted into the here and now. The words 'Hanging, 19 years old' are up there on the screen. Can I get there while they get an ambulance to the address, a converted house just outside Croydon? I'm taking my resus equipment in with me, just in case. It's a tiny bedsitter on the top floor. One of the occupants of the house, a middle-aged man, lets me in.

'Her friend came round and found her,' he half-whispers. Then he lets himself into his flat and leaves me to go up. The girl's body is out of sight, in the tiny bathroom, covered by a sheet. The friend who made the shocking discovery is sitting there in the bedsit, teary and clearly devastated by her discovery. Who wouldn't be? She looks a bit incongruous in this rather down-at-heel house of scruffy bedsits. She's pretty, smartly groomed and sports a slimline suit and high heels – office gear probably. After checking the body in the bathroom and making my calls, I go to talk to the girl. Her name is Lizzie. She's holding a crumpled letter in her hands.

As I sit down I can't help noticing a large framed picture on a small table, the only photo in the sparsely furnished room: it's of Lizzie.

'I came down this morning and found this letter on my doormat. I knew straight away it was Debbie's writing,' she sobs. 'As soon as I saw it, without the postmark, I knew what it was.'

Debbie and Lizzie have been best friends for years. Their circumstances are quite different but their bond is such that Debbie has written to her friend telling her what she is about to do. Lizzie hands me the note for me to read. It's written quite neatly and it's carefully worded, a suicide note that doesn't begin to convey the agony of the writer. Debbie says she's sorry for inflicting her sadness on Lizzie, but she can't take any more. She's too unhappy to carry on living. 'I'm ending it all, Lizzie. Probably, by the time you get this, I'll be gone,' it begins. Then it goes on to thank Lizzie for being such a fantastic friend, 'my one true friend in the whole world'.

'I couldn't quite believe it,' says Lizzie. 'I thought if I could get to her in time I could convince her she was wrong. I tried her mobile but it was switched off. So I got on a bus. I was shaking all the way here. But I didn't want to ring 999, just in case.'

When she arrived at the house one of the neighbours, leaving to go to work, let her in. Debbie's door was unlocked. And there, in the lonely bedsit, her friend found her: Debbie had been true to her word. This was no idle threat, no cry for help. She'd hung herself with her dressing-gown belt around her neck. She was hanging from the shower.

Lizzie starts crying uncontrollably again and I say what I can to comfort her. Death by hanging is a grim sight and I know it will be some time before she can erase the memory of her friend hanging there, limp, discoloured and lifeless. Probably Debbie dropped the letter off, not thinking that Lizzie would come over and discover her. She'd have imagined that her friend would first ring 999 and emergency crews would find her. Then slowly Lizzie tells me more of the story.

'We were really close right from secondary school. We met there. She'd been in a children's home when she was little. We both decided to leave school at 16 but we stayed close. I knew things were very bad for her – she'd been on anti-depressants from the doctor but she kept stopping them, saying they only made things worse. She worked in a reception job for a while, but she couldn't hold down a job. She used to say she didn't think she'd ever have a husband and kids, that sort of thing. But you don't want to believe it at the time, do you?'

Lizzie comes from a fairly comfortable background with loving parents and two younger brothers. She works as a PA in a swish ad agency in the West End. As a child she'd started out in a private primary school but had wound up at a state comprehensive when her father's business ran aground. Debbie, however, had had a dreadful life. She'd been sexually abused by her father as a young child. Then she'd been taken into care, where sexual abuse continued from one of the staff. Eventually she'd been fostered. But the damage had been done. Debbie had never felt especially loved or valued by anyone until she ran into Lizzie. Debbie was protective and stepped in when Lizzie's 'posh' accent was close to getting her a beating from some bullying classmates. So to Lizzie, Debbie was a bit of a guardian angel.

'Debbie always said she wished she was more like me,' says Lizzie sadly. 'But I couldn't understand that because Debbie seemed so tough and resilient. But she used to keep saying she was weak. She was always running herself down.'

I suspect that Lizzie, without realising it, gave Debbie something she'd never had enough of: love, acceptance and support. But tragically it wasn't going to be enough: the damage had been done to Debbie's self-esteem through the years of neglect and abuse.

'You can't blame yourself,' I tell Lizzie as the police arrive. She jumps up as I go to leave.

'I have to ring my office,' she says by way of farewell. I hear her, as I go down the stairs, explaining to her boss what's happened and why she needs today off. I know that the awful image of her friend, hanging there, will linger. I just hope it fades eventually.

Amazingly, a couple of years later I see Lizzie again briefly. She's working as a student nurse in a local hospital. We say hello but we don't talk. Maybe she doesn't want to be reminded of that awful day. But I'm glad to see that she's nursing. Maybe she's finding some comfort in helping others.

ACCIDENTS

A NICE COAT OF GLOSS

In the language of 999 services the word 'trauma' means an accident, not an illness. Usually a trauma call comes through as 'Hit by a bus' or 'Fallen from a height'. And, for some inexplicable reason, most of the trauma calls I'm sent out on involve people falling from a height. In fact many paramedics find they get job after job involving the same kind of emergency: some of my workmates wind up delivering lots of babies, others find themselves constantly called to road accidents. There's no rhyme or reason to it. It just happens that way. But I do get lots of people falling off ladders, scaffolding, buildings.

Today it's a man of 49 who has fallen from a ladder. That's all we know when two of us turn up at the house, to be greeted by his wife. He's lying there outside their front door, face down in a sea of white gloss paint.

'It was the dog,' the wife explains. Huh?

'Brian was painting the gloss on the woodwork on the first floor window and the dog knocked the bottom of the ladder.'

Brian had turned to see what was happening – and came tumbling down, a fall of about fifteen feet. The tin of gloss paint, perched precariously on top of the ladder, had gone tumbling down with him. Now it's virtually covering his entire body. You can just about see his shoes. It's gone everywhere, around his teeth, in his ears, his mouth – staining the concrete outside the front door. And he can't move his arms. But, thankfully, he's just about conscious.

'Get it out of my eyes, for fuck's sake. I think I've broken my arms,' he moans as his wife runs in and out of the house with cloths and tea towels to mop up some of the ever-spreading gloss. We try to clear his nose and eyes as best we can with the towels. But there's gloss everywhere. And the worst of it is, he's in pain. We can't attempt to move him until we do something about the pain. Lying face down with two broken arms is the worst position any trauma patient can be in. First, he's going to need a neck collar, as falling from that height could mean a neck injury. Fixing the collar would be easy if he was on his back, but this is more complicated. For starters, we need more people. We call through for another ambulance. Sorry, nothing available right now. So we contact the fire brigade for extra help.

Brian is groaning and very distressed. I give him a painkilling injection in his hand – the only bit of his body I can get to, though you can hardly even see the veins in his hand to get the needle in – as well as some gas and air. Within minutes he's easier and – magic – the fire brigade have turned up. They can help us log-roll him: keep his body in a straight line as we gradually turn him from front to back, to minimise any spine or neck damage he might have.

Then I manage to attach a splint to his arm, in case he's broken it. But after a bit he says, 'Er… aren't you doing anything about this arm, love?' Well done, Lysa, you're having a Barbie medic moment. ('Barbie medic' is my nickname among my workmates.) I've put the splint on the wrong arm! So I change it – not that easy when a person is smothered in gloss paint. We're all covered in it too. We'll definitely be needing new boots and trousers afterwards. As we finally lift the man into the ambulance, the wife comes over to get in with us.

'I've been nagging him for weeks to do this bloody job,' she says bitterly. She's probably starting to feel a bit guilty now.

'Yeah, and I didn't wanna do it, did I?' croaks Brian as the ambulance revs up. 'It's all 'er fault.'

They don't argue any more on the drive down to the hospital. In fact they don't say much at all. And it wasn't too bad: he'd broken an elbow and a collar bone. But I bet you anything you like Brian will be recounting this story to his mates down the pub for ages to come. And he's never ever going to let up on digging at her about what's happened. Typical, isn't it? Man falls off ladder, doing a job he should have done years ago. And what does he do? He blames the missus. Why can't he blame the dog?

BLOOD AND GLASS

Summer in the city. I'm just back from a terrific holiday in Lake Garda with five girls: sun, wine, the lake and lots of beautiful people. Now I've got the after-holiday blues you always seem to get after a change of scenery: why don't I live in Italy? That sort of thing. Today we've got a researcher, Ellen, a former nurse, travelling with us. Her work is to track the work of the emergency care practitioners as we go out on each call. The idea is to follow our on-the-spot decision making. We do this a couple of times a year – and sometimes it can prove useful, especially if it's a really serious job.

It's 6pm. Just four hours to go. But Ellen's finishing now. A fairly uneventful shift so far. But just as I'm thinking about dropping her off at the station, we get a bad one: 40-year-old man, arm bleeding, blood spurting from wound, coming from an artery. This is a situation where a person can bleed to death – very quickly. Ellen opts to come with me. Now the adrenalin's pumping, the siren's going full blast. It's a work-related accident and we're not far from the industrial estate where the injured man is. Someone outside flags us down and we manage to get into the estate and, amazingly, quite close to where we need to be.

The man's on a chair, next to a table, feet propped up on another chair. They've moved him from the area where the accident happened. One of his workmates is standing beside him, holding the injured arm in the air. Another one has managed to put on a hastily made, tea-towel tourniquet, twisted around the bleeding right arm. I'm a bit wary of tourniquets. They go in and out of fashion. They can completely cut off circulation to a limb – it's dangerous if they're left in place too long.

'How long's that been on for?' I quiz them, scanning the man's face, which is pasty. He's utterly drained of colour. Sweat is also pouring down his face. Yup, looks like he's in shock. And now his eyes are starting to roll back in his head.

'I'm going, I'm going,' he mumbles.

I think, Oh no, you are. It looks as if he's about to lose consciousness. On the other hand, he could just be reacting to the shock of it all, especially seeing his own blood. We are racing against the clock. I start to check his blood pressure as two of his workmates fill us in on what's gone down. Thankfully, they're sensible, straightforward types. They haven't panicked or lost it. Which makes all the difference.

'I was on the phone to my girlfriend and all of a sudden I could hear him shouting really loud,' one workmate explains. 'Even my girlfriend could hear the shouting. I just thought it was a joke at first. I told her, "Oh, it's Bob, he just wants an ambulance." I didn't realise he was shouting, "Get me a fucking ambulance now!" for real.'

Rushing in to help a very agitated Bob, he saw blood spurting everywhere from the huge wound on his arm. Bob had been carrying a big sheet of plate glass and accidentally knocked it. The glass broke and the cut edge managed to slice his forearm almost to the bone. Blood went everywhere – definitely a severed artery. A person can easily die in a matter of minutes. But luckily these two guys didn't flinch. They did everything right, calling 999, using tea towels to staunch the wound, keeping him seated and putting his feet up in another chair. Elevating someone's feet after an accident like this is exactly what you should do: it shifts the blood from the lower limbs to vital organs. They thought quickly, made the tourniquet for his arm and kept it held up. Incredibly, another workmate had looked in, seen what they were doing and insisted they were wrong.

'Don't do that, put his arm below his heart,' the man told them. This was 'advice' that could have killed Bob. Sensibly they ignored him.

'You did all the right things,' I tell them. They look relieved. But now I've checked I've found his blood pressure is really low and his pulse is racing. He's definitely gone into shock.

'Keep holding his arm up, but squeeze a bit tighter,' I say, putting a cannula into his other arm to attach a bag of fluids. Ellen, now in nursing mode, is running back to the vehicle to get a bag of fluids. Hopefully we can get them into him and bring his blood pressure up again. Once the line's attached and running through, it works a bit like a hose pipe: as you fill it, the pressure rises. Yes. It does the trick. With help from the fluids and his workmates continuing to hold his arm up, Bob's blood pressure comes up, he stops sweating and his colour returns a bit. And – the mark of a man in survival mode – he's even starting to joke about the situation. A back-up ambulance has arrived with two more girl paramedics, so he's surrounded by women, all giving him their undivided attention.

'How lucky am I, guys?' he chirps to his mates. 'Four lovely ladies!'

'Well, you could have been luckier and not sliced your arm off,' I tell him as the lovely ladies get him into the ambulance. The mood now is positive. But this is still an emergency. Tourniquets cannot stay in place for too long because all the time they're cutting off circulation. And while it's not been on for too long, every minute still counts.

Just before we race off, I manage to get a quick peek inside the room where the accident took place. Have you ever seen the inside of a slaughterhouse? Blood everywhere: floor, ceiling, walls. Welcome to the Red Room. Think of *The Texas Chainsaw Massacre* and you won't be far off. I can see the edge of the glass that cut Bob: there's a tidemark of blood on it where it's sliced into his arm, showing just how deep it went in. I estimate it's so deep there's a risk it's damaged the bone too. And it's an extremely heavy piece of glass.

In the ambulance, Bob's still pretty perky. But at the hospital, the minute the tourniquet comes off, he starts bleeding profusely again, and it has to be quickly replaced. Then he's rushed into emergency surgery to repair the damage. Our work's over. But he really was on the brink when we got there. Getting the fluids into him made a difference. But it wasn't us, it was his workmates who really saved him.

'Had that stupid guy been in charge, they'd be at his funeral the next week,' Ellen says as we head for the station. But that's typical of the job. We turn up to see someone looking ghastly and about to give up the ghost but with the right equipment – and everyone involved working together – it can be turned around. And you're left with a man cheerfully flirting with you. Does that happen in many jobs?

TRAPPED IN THE BANK

Female, trapped in a lift. I'm reminded of the time it happened to me, when I was on my way to help an elderly woman in a nursing home. The lift got stuck between floors before we could get to her and the fire brigade had to come and get us out. It took half an hour to free us. And the woman died.

I'm on my own, but the other emergency services will be alerted. The address is a local bank. Sure enough, when I get there, the fire brigade, police and an ambulance have all arrived. Then one of the firefighters briefly fills me in. One of the bank staff had been sending a safe jam-packed with bags of cash and currency down to the basement in a small service lift. For some mysterious reason, she decided to get into the lift with the safe. The safe weighs a ton and the lift isn't even designed to take people – it's not even high enough for a person to stand up in. Yet she went into the lift first and her colleagues wheeled the lift in after her, virtually trapping her against the back wall. So she was standing, bent forward, over the safe – with barely any room to move. As the lift starts to descend, the wheels at the front of the safe become trapped between the lift and the ground floor. The safe tips backwards and the lift is trapped between floors.

And the woman is trapped too, with the top of the safe crushing the lower part of her chest. It's incredibly dangerous and rescuing her will not be simple. Even the fire brigade are stumped: how can they release her without crushing her to death? The atmosphere in here is tense. The staff are milling around, shocked and quite sombre. The manager is in tears.

We can't see the girl properly. I try to lean in to talk to her but all I can see are her eyes and forehead. We manage to pass in an oxygen mask and rest it on the top of the safe to waft some oxygen to her; it's impossible to reach in to get the elastic around the back of her head. She's conscious and terrified. Who can't relate to that? I'm quite claustrophobic myself and the thought of being in a situation like this is very scary.

We can't even monitor her pulse or blood pressure as it's impossible to get any monitoring equipment on to her. All that's possible for us is to talk to her and try to keep her calm.

'Don't worry, we're all out here, looking after you,' I say soothingly. I know the safe must be cutting into her ribcage. 'Are you in pain?' I ask.

Her answer is chilling: 'It's crushing me. I can't breathe. It hurts!'

'It's all right, we'll get you out,' I say. We sound so confident, so calm. But we're fearing the worst – collapsed lungs, crushed liver or spleen. Horrific injuries.

Having tried all the obvious manoeuvres, the firefighters are still desperately trying to come up with a sensible solution. The problem is, the safe is so heavy, the slightest movement would cause the weight to shift further on to the girl's chest. Attempts to lever the front of the safe merely increase its pressure on her. And we can't give her any pain relief, as we can't get near enough to her to put in a cannula. As we all wait helplessly, the tension is palpable. It mounts as the minutes, then an hour, then two hours pass.

'Look, this is her,' says one of her workmates, showing me a picture of the girl on her bus pass, smiling and happy. She's genuinely trying to be helpful. But I'd rather she hadn't done that. Sometimes the less you know about someone the better – especially if their life is hanging by a thread.

A generator has been set up by the fire guys to power specialist cutting and lifting tools. They've tried to cut through the walls housing the lift shaft but that didn't work. The air is thick with fumes from the generator, and the noise of the generator is deafening. The front door of the bank is wide open, in an attempt to ventilate the place. Outside it's sheer chaos. The street is closed to allow all the large vehicles with specialised rescue equipment to park close to the bank. I go to sit with a bank worker who is visibly distressed and in floods of tears.

'I feel so guilty,' she sobs. 'It's all my fault.'

'Of course it's not your fault,' I tell her. 'These things happen sometimes.'

'No, it's down to me,' she tells me between sobs. 'I dared her to go in the lift.'

What can you say at a time like this? 'You complete fool, why did you do that?' or, 'Yes, it's your fault, you'll have to live with it for ever.' Instead I try reminding her how brilliant the fire team are.

'Don't worry, they'll get her out, they're doing everything they can,' I say. 'In a few days' time you'll all be laughing about this.' What I don't tell her is that the worst outcome is already being considered. The police have managed to contact the girl's relatives – and they're now making their way here.

The firefighters think they've come up with a workable solution. They can cut a hole in the safe and remove all the money inside. That should reduce the weight a little. So the cutting machines get to work making the hole and within half an hour ten of us are standing in a line, passing the bags of money along to each other. The bank looks like it's been raided: furniture pushed over to make room for the emergency teams, money bags piled up and the police standing guard over the doors.

Then another bright idea: if they cut off the front wheels of the safe too, we should be able to get her out. So they set to it. But what we don't know is what state she'll be in after over two and a half hours in the lift. So we set up the trolleybed and resus equipment within easy reach, just in case.

With the weight reduced and the wheels gone, the safe can be pulled off the girl. She collapses in a heap at the bottom of the lift. I get in, assess her breathing and give her oxygen.

She's shocked but she's breathing OK. We carefully lift her up and out and on to the trolleybed, then give her fluids and pain relief. She's tearful.

'Is my mum here? I want my mum,' she keeps saying. In fact the family have been redirected to the hospital to wait for her. The ordeal is over. She's rushed to the nearest specialist trauma unit to be treated.

Later one of the firefighters confides in me: 'This is the trickiest extrication I've done in 15 years.' So tough was this job, they'd even managed to get a crew to film the rescue operation to use for training purposes. Not that there'll be much to see – most of the footage would be of the backsides of the emergency crews as they took it in turns to reach into the shaft and reassure the poor girl.

The girl overnights in hospital but emerges virtually unscathed from her hideous experience apart from some bad bruising. She'll never forget it – and nor will the rest of her workmates. As for me, like everyone, I'm just thankful for the resourcefulness of the fire brigade. The girl was a hair's breadth from being crushed to death. But it didn't do much to help my fear of lifts and enclosed spaces. Can you blame me, really?

BELT UP

One of my closest friends is a firefighter who, after witnessing many road traffic accidents over the years, says the nature of people's injuries has changed for the better. Years ago, before seat belts were compulsory, he says it was quite common to turn up to a fairly low speed collision and find a person slumped over the bonnet in a terrible condition. Sometimes they'd be sitting back in the seat having hit the windscreen, with hair, skin and blood trapped in the shattered glass. They would often have shocking facial injuries as well as dreadful damage to the rest of the body.

Thankfully cars are now much safer. There are airbags, crumple zones and seatbelts. Being involved in a road accident at less than 30 miles an hour now usually means nothing more than a bit of whiplash for the occupants of the car. We frequently arrive to find everyone is out of their cars, exchanging details and ringing friends and relatives on their mobiles – often too busy to talk to us! Children in car seats come off best of all. These seats are fantastic safety measures.

But try as I may not to get preachy or climb on my soapbox to sound off too much, one kind of motorist that makes me fume with volcanic rage is the driver with a child – or many children – sitting or standing in the back of the car. The adult is selfishly and safely strapped in with no regard for the safety of their precious tiny passengers. Or you see a belted-in adult with a child on their lap, either under the belt or on top of it, both positions horrifically dangerous. Don't they realise that the child between them and the seatbelt could get crushed in an accident? Or that the child sitting on top of the seat belt could be thrown from their grasp in the event of a sudden collision?

Even with a relatively minor collision an unrestrained child is bound to be catapulted through the air – only stopping when their little body hits something hard and solid. Occasionally they'll be thrown clear of the car. And we then have to go and look for them, lying unconscious in a hedge, ditch, gutter or someone's front garden.

When something bad happens to a child in a road accident and the parent has clearly ignored sensible safety precautions and let their child travel without a belt or kiddy seat, it's a real struggle for me to successfully maintain a professional, reassuring facade when the child is seriously injured – or worse. My emotions are ready to take over. I can keep them in check – just.

I've arrived first at a traffic accident at 9am. It's a fairly busy back street used as a bit of a rat run by the parents and childminders who have just dropped their children off at the nearby schools. One call has come up as 'Damage only', but there's been more than one caller. Another suggested that someone could be badly injured. An ambulance is also on its way but as I drive to the scene it's heartening to see that the two cars involved seem to be only minimally damaged. It looks as if one has gone into the driver side of the other as it pulled out of a side turning. In one car there's a woman still sitting in the driver's seat, chatting on her mobile. Sitting on a wall near the other car is a young woman holding a child. The child looks about three years old and is quietly sleeping in her mother's arms with her long, blonde hair covering her face. Good. Everyone looks uninjured and this could be a pretty straightforward call for us.

When I introduce myself to the mother she looks at me and says, 'We're fine, we're fine. No thanks to that bloody stupid cow who pulled out without looking. She was on her bloody phone, y'know!' she adds, rocking the child. Then a young man walks towards me. He's standing there, hovering, obviously eager to tell me something. I walk up to him, leaving the woman on the wall.

'The kid was in the footwell,' he says.

'Sorry?' I say, not quite understanding.

Then he tells me. He was walking past, saw the accident. The impact made quite a bang. He ran over to help. Like me, he initially imagined that no one had been badly injured. Until the mother started to cry: 'My baby! Where's my baby?'

The man looked around at the back of the car – nowhere could he see a child.

'I thought she was just shocked and confused,' he tells me.

Then he heard a whimpering sound from the passenger-side footwell. And there, underneath a bag and partly covered by coats, was the injured child, curled up and barely conscious. She'd been sitting unrestrained in the back seat. The force of the impact threw her up in the air and into the footwell.

I think, Christ almighty! This changes things quite a bit. Now I'm looking at the child and the scenario becomes a real emergency. And she's not asleep at all. She's barely conscious. In a situation like this, without back-up, you need to think on your feet – double fast. I get the mother to stay as still as she can to support her child while I check that she's still breathing. I get oxygen on to her little face while trying not to move her at all, then I pull out the phone to ask for police – and find out when back-up is due to arrive. The helicopter is also requested, but unfortunately it can't get there this time. I'm really worried now. The little girl, Jess, may have spinal or internal injuries.

When the ambulance arrives, we need only move her once – and we can then do it with many hands on deck. Within five minutes the welcome sound of the screaming sirens signals that they're here. Teamwork: we quickly manage to package the little girl up for the trip to hospital, getting a collar secured round her neck and strapping her to a spinal board. As we do this we notice that she's a reddish, mottled colour across her abdomen, a sign that there may be internal injuries. She could be needing urgent surgery. She looks so tiny lying there. It tugs at everyone's heartstrings to see her pale face, hear her tiny whimpers. It's now obvious she's in a bad way – a well child of that age would be having a hissy fit at the indignity of the collar and spinal board and crying for their mum.

And the mother? The passer-by who witnessed it all and then rang 999 tells me that the mother became quite hysterical when her child was discovered in the footwell. Then she'd thoughtlessly grabbed her and gone to sit on the wall, waiting for help, cursing the other driver over and over again, insisting, 'We're fine, we're fine.' And she continues to heap abuse at the other driver as we rush, sirens at scream level, to get her little girl to hospital.

'She was on the fucking phone. She was on the phone!' she keeps repeating angrily.

I say very little to her, other than a few words of reassurance that her daughter will be looked after once the hospital takes over – and yes, the woman was wrong to be on her mobile. But had this woman used a child seat for her daughter, the injured little girl would almost certainly be unscathed right now. It's a 'bite your tongue, Lysa' moment. And it gets worse. The doctors discover there is internal bleeding from the little girl's liver. Her body struck the seat in front of her as she flew into the footwell. Despite surgery and the best efforts of the paediatric intensive care team, I'm told later that she has died.

I never saw the mother again and I've no idea if the police decided to prosecute. All I know is, if you take nothing else away from this book, please always take care to strap your tiny passengers into the appropriate car seat. If your child refuses to wear the seat belt, pull over and don't drive – you're in charge, after all.

Time and again you hear parents saying, 'Oh little Johnny won't wear it, he screams. What can I do?' Well, here's what you do. You get a grip. Or you may be burying little Johnny. Or pushing him around in a wheelchair until he's Big Johnny and still needs your help to wash and toilet himself.

MR GOBBY

The 999 crew tends to attract an audience, no question of that. If the police turn up around the same time as us, they'll deal with the crowd of bystanders or nosy neighbours as it gathers. But without their help, people in the crowd can often cause problems for us. They can even hinder us – especially the stroppy ones who've got nothing else to do but cause trouble.

I was among a number of paramedics to be filmed for a TV documentary. On the last day of filming we're called out to a road accident, a seven-year-old struck by a car, resulting in 'serious mechanism injury'. That means trouble. What we don't yet know is whether there's been high impact, meaning the child has been thrown up into the air and landed in the road. If that does happen it could mean a second injury following on from the bash from the car. And, like all calls involving kids, it might be serious because children can deteriorate very quickly. Quite often they don't show the same signs of shock as an adult until late in the game. And then it can be a really sudden deterioration that's impossible to reverse.

En route I explain to Andy, the producer cameraman accompanying me, just how serious this call might be. 'We might need the helicopter,' I warn him. Andy's sensitive enough to realise that he might have to stop filming.

'To be honest, I don't think I should film it,' he says. 'It doesn't feel right, filming a kid hit by a car.' I'm relieved. And if it is bad, Andy might be able to help.

When we arrive there's a big crowd standing around a young boy. He's sitting there in the middle of the road crying and visibly distressed. But he's conscious and not lying flat on the ground, which is positive. There's no visible injury. Yet he's quite hysterical with shock. I need to know more about what's happened. Has he been bounced off the car? If he has, despite looking OK, he might have internal injuries. Or his spine might be damaged. As for the car driver, she's still sitting behind the steering wheel, bawling her eyes out. And despite the rubbernecking crowd, no one is making any attempt to talk directly to the boy.

Then someone says, 'That's the mother,' and points me to a well-dressed lady in her thirties, talking furiously into her mobile but, strangely, doing nothing to comfort her distressed son. She could be a total stranger for all the attention she's paying to him. It's peculiar. Most people in this situation would stop talking on the mobile to acknowledge the presence of a paramedic. Not this woman. That's bad enough. But there's another distraction. A scruffy, tubby, twenty-something man in a baseball cap, trainers and jeans who is definitely under the influence of something. He's a real aggro type – the sort of bloke you only have to look at to know he wants trouble. I'm trying to help the boy and am attempting to hold his head still in case there's a neck injury, but the troublemaker homes in on Andy, who isn't far behind me. He's spotted Andy's camera.

'What the fuck are you, some kind of bleeding paedophile?' he leers at Andy. 'Why've you got a camera when you're supposed to be helping the kid?'

'No, mate, the camera's off, we're not filming, just calm down,' says Andy, visibly shaken by the man's aggression. But Mr Gobby won't be appeased. He continues to rant and rave while I'm trying to get the boy to tell me what's happened. I get nowhere. The boy just screams and screams, shouting in a foreign language, waving his arms towards his mum. But she's still ignoring him, totally absorbed in her conversation.

I attempt to soothe him: 'Mum's here, just tell me what happened.' But it doesn't work. He's just screaming. His mum's on Planet Mobile. And Mr Gobby's virtually foaming at the mouth.

'Why've they sent a bloody woman?' he spits, edging a bit too close to us, determined to create the maximum amount of aggro.

Andy does his best to calm Mr Gobby down: 'Look, mate, she knows her job, just let her get on with it,' but we can both see how close we are to watching Mr Gobby throwing a punch. Naturally no one in the crowd does a thing to help the situation. They just stand there, silently staring, while the mother continues to talk, rapid fire, into the mobile.

'Please tell me what's happened!' I say to the mother, cradling the boy as best I can – and getting increasingly worried. If he does have internal injuries, every minute will count. Incredibly, she just waves me away with her hand. This is really awful. No one can tell me what's happened – and a total stranger wants a punch-up – or worse.

'You leave 'er alone and get on wiv your effin' job!' Mr Gobby yells at me.

'How can I if I'm arguing with you?' I counter, but inside I'm getting more and more concerned. We need extra help and we need it fast. I get the mobile out. Around me is total chaos. We're in the middle of the road, traffic's piling up, drivers are blasting their horns in a noisy chorus and the bystanders continue to stare, doing nothing. Do they think they're watching the telly? Is real life too much for them to handle? Then my mobile rings.

Still holding the boy's head – now I'm clutching it to my tummy – I manage to answer it. Wrong number!

'Don't ring this number again!' I plead and at last get through to the control room to ask for police assistance – fast. In the meantime Mr Gobby has started World War III with the car driver.

'You were speeding! I saw you! It's your bloody fault!'

Stupidly, the driver is bothering to argue with him. 'No! I swear I wasn't!'

And still the mum is talking into the mobile! The language

she's speaking is not recognisable. But whoever is on the other end, they seem far more important than her injured son in the road. Andy's gone to the vehicle to get a blanket and I'm attempting to stop the boy moving. Then, without warning, he stands up, points to his leg and starts screaming again with renewed vigour.

'For God's sake, put that phone down and talk to me!' I yell at the mum. She looks over. Oh, wow, she's decided to communicate with me.

'I'm talking to my husband!' she tells me.

Of course. Silly me. Your son is lying in the road, possibly badly injured. But hubbytalk comes first. Andy, meanwhile, is helping me now, rolling up the boy's trouser leg and telling me what he can see as I hold on to the boy, trying to keep his head and neck still. There's a deep cut in his leg. But it doesn't seem as if any bones are broken.

'Good. The police'll be here any minute,' I say quite loudly, hoping fervently that Mr Gobby will hear. And guess what? It does the trick. He vanishes into the crowd. Simultaneously the mother finishes her call and decides to fill me in. She's quite polite – and calm. They were crossing the road close to the busy junction and the boy was a step in front of her. The car, already slowing down, somehow managed to hit the boy – and he fell right in front of it. Thankfully, he didn't go under the wheels, nor was he pushed up in the air or bounced off the car. Looks like this kid is one of the lucky ones – the leg injury could be it.

Now the police arrive, control the traffic and make it clear to the crowd, who've now got their mobiles out and are clicking away, that today's free entertainment is now over, it's time to get back to their own uneventful lives. Or the telly. Reluctantly they disperse. Nowadays the mobile-phone photographers are often part and parcel of a road accident. But the ghoulish element is troubling. You get a tragic situation where someone is run over by a lorry and killed, yet people will line up taking photos of a blanket. Usually it takes only one person to whip out the mobile and do it and the rest will follow suit.

'How would you like it if that was you lying there?' I often want to ask them. This senseless voyeurism is scary. Unless you're going to be useful, walk on.

Now an ambulance crew has arrived. The boy has calmed down and is clutching his mum's hand. I apologise to her for yelling. 'All I could see was you on the phone but no one could tell me anything.' She's very gracious. She understands, she says.

'But you see, I had to tell my husband.' Clearly, had she not done this, chapter and verse, there'd have been a problem. The domestic chain of command was such that telling him took priority over reassuring her screaming son.

Later Andy and I debrief. The boy's injury is relatively minor.

'A pity we didn't film it,' muses Andy. 'It'd have been a good way to make the public see what the emergency services have to deal with, especially if the paramedic is on their own.' He's so right. The potential for tragedy was there. If Mr Gobby had resorted to violence and the boy had had bad internal injuries, it could have been a disaster, given the ignorance of some people – and, I suppose, other people's marital arrangements.

BRICKS

Call-outs to building sites are not uncommon. After all, they're pretty dangerous places to work. But today's call sounds nasty: seventeen-year-old male trapped by collapsed wall. Unconscious?'

I'm in the car I usually work in but it isn't equipped for a situation like this – no trolleybed, for instance – so Chris, who is working with me today to observe and learn, gets on the phone to find out if the helicopter is available. Police, ambulance and fire brigade are already on their way.

At the site in Croydon we're greeted by a very anxious-looking site manager: he's really shaken. Something tells me he's never witnessed an accident quite as bad as this. And once we've donned our hard hats and reached the scene, we're a bit taken aback too. A fifty-foot-long wall has collapsed, without warning, on top of a young workman. Fortunately he was walking away from it when it collapsed, so he was on the periphery of the impact. But he's been flung to the ground, face down. And he's been covered by the resulting avalanche of bricks and rubble. By the time we arrive, his workmates have managed to remove a lot of the bricks, though some still seem to be underneath his body. Bricks, rubble and dust everywhere, a toppling mountain of bricks – and this guy's in the middle of it. Awful. He could easily have been suffocated.

Apparently he'd been unconscious at first, but by the time we get there, he's come to. Looks like he's breathing OK. It's a cold day and a very noisy site. I kneel down next to him and ask his name.

'It's Paul,' he says. So far so good.

'How's your breathing, Paul?'

'It's OK, but when I take a deep breath, it hurts.'

I use my stethoscope on his back and manage to listen to his breathing. He is also complaining about back pain. He's likely to have injured his spine. Gently feeling down his back, it's obvious that the pain is quite bad around his lower back. Oh, how I hope it's not spinal damage.

'Looks like we will need the helicopter,' I tell Chris, who gets on the phone to organise it. The chopper will bring a doctor and they'll be able to take him directly to a trauma centre, probably the Royal London in Whitechapel, where there's a landing pad on the roof. Now an ambulance team has arrived with more hands to help us. As we check blood pressure and pulse and start putting a collar on him, all the things we need to do before getting him on to a spinal board, he's chatting quite normally to the other paramedics. But you can see he doesn't want to move at all for the pain. A tourniquet goes tight round his arm and a cannula for fluids through goes into his elbow. It's not the easiest thing to do if someone's lying face down. He gets morphine too through the cannula. We can see he's in a lot of pain, though he's being manfully brave about it all.

But the morphine is a big help. Within seconds you can see he's easier. He's conscious, which is good, but he's had a head injury and been unconscious, which isn't so good. There's a chance of a bleed into the brain. The other thing worrying me is his neck. We can't tell right here, of course, but he could also have a serious neck injury. But at least we can move him now the crew are here – you need four or five people to do a log-roll. One person takes the head and everyone else goes along the person and gently takes a bit of their body, in order to protect the spine from any further damage.

But then we decide not to move him until the helicopter arrives, just in case. If there is any further damage, there's a chance he could wind up paralysed from the neck down. He's only seventeen. It's a terrifying possibility which we have to consider.

Now we're starting to cut his clothes off so we can get a better look at his injuries, replacing his clothes with blankets because it's so cold. His mates are standing there watching us, not saying much but silently rooting for him. They had the common sense to pile their coats over him before we arrived. Any one of them could have been standing there when the wall collapsed – you can almost hear them all thinking this.

One, an older guy, speaks up for all of them, 'Can you tell us what the score is?'

All I can say is we don't yet know what his injuries are. I don't say it out loud but I know the consequence of this type of accident could go either way. It's not just about survival. If Paul has a spinal injury, it may eventually mean life in a wheelchair and loss of many functions. Or he may be bleeding inside his head, which can lead to brain damage and another kind of hell. How bad are these prospects for anyone, let alone a teenage boy?

Just then the loud noise of the Helicopter Emergency Service breaks into my thoughts. The pilot is amazingly adept at landing anywhere – and within minutes we can see the team, two doctors and a paramedic in orange boiler suits, running towards us with their heavy equipment. They've already radioed ahead to the police to tell them where they're landing, so the police have been able to give them a lift to the site.

One of the paramedics is a friend, Jack. We briefly exchange hellos and start the log-roll to get the boy on to the board, then the doctor has a good look at him to double-check what we've already told him. Is Paul now ready to be airlifted? But there's no airlift, after all. The doctor doesn't think he needs to go by helicopter. He says Paul's safe enough to go to the nearest trauma centre by ambulance.

Phew. Good news already. His mates brighten up visibly too. We all help get him into the ambulance and everyone gets a chance to say goodbye – and good luck. Paul's shaken and pale but knowing he's not bad enough to be airlifted is a small boost to his spirits. He looks a lot more comfortable. Turns out he's got a wife – at seventeen! – and he wants her to be contacted so she can come and find him at the hospital. How close she's come to being a young widow, I reflect.

Three weeks later and I want to know more, so I drive by the building site. The manager recognises me – and there's terrific news. Paul broke a bone in his back but that's it. He's doing well.

'He wants to get back to work in a few months' time,' his boss tells me. What a great outcome after all that. You wouldn't normally expect just a broken bone after this kind of accident. Had he been standing in the wrong place, the bricks would have probably killed him. Or left him without much of a quality of life. And when I mull over it later, I can see he owes his workmates a drink or two. They didn't move him, kept him still – and warm. Little things, you might think. But sometimes it's the little things that people do before we can get there that can make the difference between tragedy and a happy ending.

FAMILY VALUES

A HAPPY ENDING

'Five-month-old baby girl, floppy.'

A call like this, you don't talk much or speculate with whoever's working with you. You just get there. Who wants to see a poorly child? Or worse? Adele drives like the wind, we're not more than half a mile away. As we approach the lift we hear a male voice bawling from an open window, 'Where's that fucking ambulance?'

We look at each other. This is a serious call. And there's one very angry man upstairs. The woman on the shabby sofa has a baby in her arms. She's painfully thin and looks exhausted by life, though she can't be more than early twenties. She's sobbing, 'My baby, my baby,' over and over again. The man stands there in the corner, a beefy, hairy monster in a check shirt, arms folded, anger oozing from every pore. Can this be the dad? He must be. Yet despite his bile, he looks so detached, so uninvolved. And he's not making any attempt to comfort or console the woman.

'Give me the baby,' I say and she hands it over. Oh no. It looks like she's already dead. Her tiny mouth is open, but her body is motionless, drained of all colour, tiny arms hanging down, head hanging back. And she's so small – almost the size of a newborn, rather than five months old.

'Right, let's go!' I say to Adele. She takes the stairs, running, I press the lift button. In the lift I give the baby two little puffs of air. I rub her tiny chest, hoping desperately for a small sign of life. A miracle, please, tonight, now. Thank you, God. The baby takes a tiny gasp. Then I rub her chest again. Another small gasp. I keep going until the lift doors open. Adele has already opened the ambulance doors. Once inside, we get the oxygen out and start to waft it over the baby's nose and mouth, continuing to rub her chest, to try to help her breathe more. And yes, she's starting to take tiny breaths now, all on her own. She's even making a small mewing sound – for us the sweetest sound in the world.

The mother has followed us down. She's got a friend with her, someone we didn't even notice in the flat in our rush to get the baby to the ambulance. The mother is still hysterical, inconsolable. The friend has an arm round her, trying her best to support her.

'She's breathing but we've got to get her to hospital quickly,' I tell them. 'What happened?' The mum just continues to sob. She really is too hysterical to speak normally.

'I dunno what happened,' shrugs the friend. 'She just phoned me to say the baby wasn't well. So I went round and the baby looked terrible. So we called you.'

'Was that the father upstairs?' I ask, wondering why neither parent had called 999.

'Yeah. He was out when I got here. He must've come in just before you arrived.' It sounds plausible but there's no time to delve further. I jump back in and we're off. As we pull away, I notice some blue lights.

'The police just pulled in as we left,' notes Adele, switching on our lights and sirens.

By the time we reach the hospital and the waiting resus team, the baby is breathing fairly well for herself and even crying a bit, another good sign. But there's no question that she's in a bad way. We impress on staff how bad she's been, how we found her half-dead. Afterwards we discover that she's so tiny because she was born three months prematurely. The fact is, Adele and I could do our paperwork and make ourselves available for the next call, but we just don't want to leave her. We're willing her to pull through somehow. The mother and friend have been directed to the relatives' room. The mum is too hysterical to stay in resus, watching the team work.

Thirty minutes on and the signs are that the little girl will survive. But when we finally do come out of resus, we're confronted by three concerned coppers, wanting to know if we've just brought the little girl in here. We nod.

'What's the score, what's been going on?' says one of them.

'We really don't know, just a very poorly little girl,' says Adele.

'Any bruises?' he says.

'No, didn't see any bruises. But she was filthy dirty,

underweight, really skinny. And the nappy was heavily soiled,' I say.

'Anything else?'

'Well,' I hesitate. Then I decide to go with instinct born of experience. 'She could have been shaken. But we can't say for sure.'

They thank us and walk off. 'Doesn't look like they're going to leave it like that, does it?' says Adele.

A few days later I'm at the hospital and ask for some feedback. The baby has had a CT scan. There is evidence that she's been shaken, probably on several occasions. There's some sign of bleeding in different parts of the brain.

'She's probably going to have some brain damage,' I tell Adele over the phone that night. 'That baby's had a terrible life.'

Weeks later there's more info. The little girl is doing really well. Then, a month or two later, we find out that the whole thing has gone to court. The baby has been taken into care and then adopted. From what we learn, in court the mum tries to pin the blame on the father and the father blames the mum for shaking the baby. In the end, they're both charged with neglect. At least, though, the baby now has a chance in life, being adopted by a family who will love her.

But the funny thing is, hundreds of calls on, I can never really stop thinking about that little girl, so tiny, so determined to survive, despite her brutal start in life. For years she stays in my mind: I wonder how she's fared, how the brain damage has affected her. And then something incredible happens.

I'm doing my training to move up to ECP (Emergency

Care Practitioner) level and as part of the training course I have to do hospital outpatient work involving children. A consultant paediatrician is going to let me spend a day with her in an outpatient clinic, so I can watch and learn more about this kind of work.

It's fascinating; I love working with children anyway and the morning goes quickly as we see her young patients. Then the consultant picks up some notes and remarks, 'This next one is a very interesting child, Lysa. Then she mentions the little girl's first name, an unusual one which I've never ever forgotten. No. It can't be! I look at the notes, double-check the age. It is, I'm sure it is.

'This child was abused as a small baby. There was some evidence of brain damage but she's made an amazing recovery and has now been adopted by a very caring family; she's a delightful little girl.' I gulp. What fantastic news. I'm almost speechless with delight.

'I think I know her,' I manage to get out. 'We brought her in here nearly dead.' Then I tell the consultant the story. She smiles.

'Well, you're going to meet her now.' I feel quite emotional when she walks into the room with her mum. And, oh, she's lovely, white blonde hair, a pretty little six-year-old, all smiles and laughter. And the brain damage isn't severe. Her IQ will never be high and she might have memory problems or be difficult to reason with. But she'll grow up, get a job, have a life, a family – a normal future. The consultant and I have agreed that we won't mention anything about my involvement in her past to the adoptive mother. And, on talking to her, it's obvious that she loves the child as if she were her own.

I sit and chat with the little girl. She's clutching a colouring book and pencils and wants me to help her. I talk to her, watch her carefully choose the colours, an adorable, ordinary little girl who has survived the very worst that life can throw at a tiny baby – and I marvel at her. In fact I want to dance round the room for joy. It's a great moment.

Later I ring Adele to tell her the amazing news. She's pretty chuffed too: what an incredible coincidence. 'You know what, Lysa, it was meant to happen', she says, ever the wise owl. 'I reckon it's nature's way of reminding us that our job's worth doing.'

Perhaps. Or is it just plain luck that gives one child the chance of a happy life, while another life may be pointlessly, needlessly lost? In a life-or-death drama families look at us, rushing in with our equipment, as if we hold the key to it all. But we don't. One thing this job teaches you is that nature works in a random, mysterious way, that's for sure. Against all the odds, every now and again the frailest person will survive, the gravely injured will come through, the old lady dying in the hospital bed will walk out of the hospital unaided. The whys and wherefores are not for me to debate. I'm just grateful good things happen sometimes.

A SCHOOLGIRL

It seems fairly innocuous. A thirteen-year-old girl has fainted, not that unusual, given that young girls have a tendency to do this. But there are times when the most routine or everyday call will prove to be something completely different.

A small, clean and neat house in a south London housing estate. Pete and I meet an elderly woman. Her name is Edie.

'She's on the sofa,' Edie says. 'She did faint but she's coming round now.'

Is Edie related to the girl? 'Oh, no,' she says. 'I don't know her. She lives a few doors away from us; we don't know her at all.' We check out the girl. Her name is Amy, and she's crashed out in the lounge. She's wearing her school uniform.

I kneel beside her. 'How do you feel, Amy?'

'Fine,' she says warily. Pete and I are not so sure. Edie then fills us in on how Amy has got here. She'd been at home watching daytime TV when her doorbell rang unexpectedly.

On opening the door she was taken aback to see a dazed and confused schoolgirl there on the doorstep, being supported by two young men.

'I'd never seen them before and the girl, well, I only know her by sight. They said they'd found her collapsed on the street, they thought she'd fainted, so they'd knocked on my door for help because my place was the nearest,' says Edie. 'They said it was lucky they were passing by.'

The two men plopped a semi-conscious Amy down on the sofa and left very quickly, leaving no names, details or information whatsoever. It's clear that Edie's completely overwhelmed by what's gone down. And she's as confused as we are as to what has really happened. We're not sure Amy has just fainted; as paramedics we can sense that something's not right.

'I kept asking her what happened but I couldn't get much sense out of her, just her name. So I rang 999.'

Odd. Amy now seems coherent enough to speak. But she's distinctly reluctant to say much. She avoids eye contact with us and just picks at her nails. Can she remember what happened before she fainted? Pete asks her.

'No.'

Has she ever fainted before?

'Dunno, don't think so,' she says cagily.

Slowly I glean a tiny bit of information. She's healthy. She doesn't take any medication for asthma or any other complaint. Then I notice the pupils of her eyes. They're strangely small: pinpoint pupils, we call them. I get Pete to double-check them with a pen torch. He looks, sees how small they are – and we exchange knowing glances. Drugs. But what kind? And how much has she taken?

'Do you take any painkillers for anything, Amy?'

'No, I didn't take nothin',' she says sullenly. Oh, yes you did, chorus Pete and I silently. It's all very mysterious. But with kids and drugs you sometimes have to be patient, keep trying to get the truth out of them. She says she feels a bit sick, but that's all. She feels much better than she did before, she assures us. We leave the bemused Edie and take Amy out to the ambulance. Once we're on our way, I start to chat to Amy. I tell her quite pleasantly that I'm concerned that she might be hiding something from us. My worry, I say, is that she might have taken an overdose of something, like painkillers, and be too scared or worried about the repercussions to tell us. And that could be dangerous for her.

At this, Amy laughs. 'I didn't take painkillers, honestly,' she says.

'Well, could you have taken something without realising it?'

Amy lapses into an uncomfortable silence. Now I'm getting somewhere. Sure enough, she then starts to tell me her shocking story. That morning she'd been skipping off school when she ran into these two young men. She didn't know them really, just enough to say, 'Hi,' if she saw them in the area.

But today they stopped to chat with her. 'Why don't you come over to our place for a pizza?' one suggests. 'It's not far.' Amy, innocently thinking the men to be kind and generous, goes with them. We only find this out afterwards, but Amy's a vulnerable kid who's been having difficulties at home with her stepfather. She often runs out of her home, roams the streets and doesn't return for hours. Her attendance at school is patchy. So she's a perfect target, wandering aimlessly around the area for these two guys who are constantly on the lookout for girls like Amy. At their flat they've given her some vodka to drink. But then she starts to feel uncomfortable. She wants to leave but these are two big guys. And they've locked the door.

Starting to feel intimidated and fearful, she sips the vodka, unsure of herself, not wanting to feel ungrateful. Then the men pounce. They grab her, put a tourniquet tightly around her arm – and push something into her vein through a needle. She slumps back, unconscious. For these men are despicable local drug dealers. And the drug they've injected is heroin. Their aim is to get her addicted – so she'll come back to them, time and time again, for a fix. A regular customer, trapped in a deadly spiral.

But it all went wrong. They have misjudged the amount, easy enough to do with someone that young who has never had the drug before. Even when it's just a small quantity, a youngster is highly susceptible to the effects of the heroin. And no one ever knows for sure the exact quality of the heroin – or what it has been mixed with before it goes on sale on the street.

For all the two guys know, Amy might never wake up. So they panic. They don't want a dead schoolgirl on their hands in their flat, their base for dealing in the drug. And they're certainly not going to call 999. So they carry her out of the flat and knock on the unwitting Edie's door. They spin their story about Amy fainting. They dump her. Even if she'd been dead, they wouldn't have cared less – or had any remorse.

Amy, of course, is streetwise enough to know what the needle contained. But I can see she really has no idea how close she's come to an untimely end. As I go to take her blood pressure I find the small red mark on the inside of her elbow where the needle has entered her skin. Then I do my best to explain that drugs like heroin can stop you breathing completely – even if just a little too much is given. Worse still, if you fall unconscious and then vomit – also quite common with a heroin overdose – you can't protect your own airway and the vomit can enter your lungs and block them. This stops you from being able to take oxygen into your lungs effectively – and it's a truly horrible death.

She listens to me intently, this vulnerable child who so nearly met a dreadful end, and I can see that some of it, at least, has sunk in: she's been lucky. But as we take her into the hospital and Pete and I hand her over to be fully checked out, I can't just feel a sense of relief that she's had such a narrow escape – I worry about her future.

We'll write down our concerns, which will eventually be passed on to social services, but who knows what fate holds for Amy? If those two guys had succeeded in their vile plan, she'd probably have been hooked, maybe led into prostitution, even prison – and one more life would be ruined. My friend Ruth, who works in the prison service, insists that hard drugs are the scourge of humanity. I never used to think too much about it. But, after today, I can see exactly what she means.

MOTHER AND CHILD

The door opens slowly, to reveal a little girl, no more than three years old. She smiles at me and I think of the past, the era of Charles Dickens, and a tiny waif adrift in the chaos of poverty in London. She's barefoot with dirt visible between her toes, wearing just knickers and a t-shirt. But she's beautiful, like a little doll with long, dark hair and a fringe.

Bob and I don't really know why we're here at a house in a rundown estate. Sometimes calls come through without much information: Mum not very well, is all we got. The little girl leads us upstairs. Her mother is in bed, young, mid-twenties, skinny. And surrounded by squalor. The place is freezing – it feels as if it hasn't been heated for years. Flick a light switch: no electricity – cut off? The bedroom is grubby, unwashed sheets, fetid air, windows never open. Next to the bed is a plastic washing-up bowl with some vomit in it. Everything spells out neglect. This woman hasn't been looking after herself – or her child. I take her arm. She feels hot, feverish. What's happened? She says she's been sick time and time again, she feels terrible. It's been like this for nearly three days.

'There's nothing else to bring up,' she tells me. 'I had to crawl to the toilet, I felt so awful.' Her eyes are sunken, her mouth is dry. And her pulse is racing: the force of her pulse is hard to follow: probably low blood pressure. Could be a stomach bug perhaps. Has she been eating the same food as her daughter, who seems OK? Yes, they've been eating the same food. I don't say so, but it doesn't look like the kid has eaten for two days either. I run through the list of suspect foods that might have caused a bad tummy upset. Chicken? Shellfish? Reheated rice? Has she eaten any of these?

'No, but my hips hurt. I ache everywhere and I've got this headache.'

'Could be gastroenteritis,' I say to Bob. A high temperature can make your joints ache and a headache can be caused by dehydration. Bob does her blood pressure, which is, as we suspect, very low. She needs hospital.

'So who can look after your little girl for you?' I ask.

'There's no one,' replies the girl. 'My mum won't do it.'

'Well, how about if we ask her for you?' I say brightly, knowing that usually we might have to talk a grandmother round – but they tend to help out if there's no other option. The girl says nothing. She's weak, exhausted. The little girl sits on the bed, watching us carefully, her finger in her mouth. I go downstairs to find the phone. Even a cursory glance tells you the house is shockingly unsafe for a tiny tot. No stair gate, nothing at all in the fridge, just mouldy bread and a tub of marge, no food in the cupboards. The bathroom reveals a tub full of scummy bath water, razor blades and contraceptive pill packets scattered around. Amazingly, next to the phone, I find the grandmother's number, scrawled on a bit of paper. I dial the number, talk to the grandmother, go through my usual spiel. But I'm taken aback by what I hear.

'No way am I coming there. You can tell her she's had her chances,' is the response. I don't know what the problem is and I can't ask her. I try again. But she is adamant: she will not get involved in any way.

'She won't change the way she lives. I did my best, I worry about them, I've cried all night. That child shouldn't be running around with the people she lets into that house. But I'm staying out of it. I'm sorry, miss, if that makes me sound cruel. But that's it.' I try to convince her, if only for the child's sake. I explain that without her help the child could be taken into care. I don't say anything about what I've seen in the house. But her response tells me she knows how it is.

'It might be better for her if they did. Then something'll be done,' she snaps before slamming down the phone. I'm flummoxed. Neighbours often go the extra mile to help, but this was a grandmother – who'd clearly reached the end of her tether. Upstairs Bob has put a line in to give the girl fluids. I don't say much to him. But in my mind, I can't help wondering about 'those people in the house'. What is this child being exposed to? Prostitution? Drugs? Both? I tell the girl, as tactfully as I can, that her mother won't be coming. And she's starting to cry.

'They'll take her away if I go to hospital, won't they?'

That is the stark reality. As I try to reassure her that her daughter will be safe, she starts to moan with pain. Her stomach is really hurting now, she says. A few more questions. Could she be pregnant? No, she had her last period a few weeks back. Did she use tampons? Yes.

'Can you be sure you took the last one out?' I ask, suddenly clicking on to what this probably is. She can't say, she doesn't know.

Everything now adds up to toxic shock, a very nasty, life-threatening infection in the bloodstream. All the signs are there: low blood pressure, fast pulse, very hot. If I'm right, it's a serious infection in her system, overwhelming her body. She needs to be given powerful antibiotics in hospital quickly. She could wind up in intensive care. Or dead.

I talk to Bob and now we're moving fast. Hurriedly we find some clothes and shoes for the little girl, get them on her and move mum and child into the ambulance and on the way to A&E. The mother is quiet. She seems resigned to her fate. The girl livens up, chatting away. She loves the ambulance, wants to know what everything is.

'What's that?' she says, pointing to a suction machine we use to clear an airway.

'That's for sucking up sick,' I tell her. She pulls a face. It's heartbreaking.

She's a little sweetheart, exposed to God knows what by a mother that doesn't know how to cope with her own life, let alone care for a child – even though she obviously loves her.

And yes, it is toxic shock from an unremoved tampon. She just forgot to remove it. It's not uncommon. Some women even pop one in without removing the old one. Yet the oversight – or carelessness – has come pretty close to killing her. But she survives, after a stint in intensive care.

As for the little girl, she eventually goes into temporary foster care. Beyond that, I don't know. Another 24 hours or so in the house without calling us and she'd have been marooned in that neglected house, hungry, bewildered – and orphaned.

This isn't the first time I've gone into a home where a child is being neglected because of the mother's chaotic lifestyle. Some people just fall through the system – because no one picks up the signs. Yes, the grandmother could have reported it to the authorities – but how could she willingly do that to her own daughter? So they stayed there, just under the radar.

It is a part of our job to report what we see if we do spot the danger signals with children. It's a statutory responsibility: we have no choice but to protect children. And two children a week die at the hands of their own parents. So the tragedy is that usually we're protecting kids from their own parents.

THE POSH HOUSE

Two teenage girls, alcohol abuse. One unconscious. The location is a large, detached house in an affluent area. Jack, a boy of about fourteen, comes to the door. His parents, he says nervously, have gone out for dinner. Earlier that evening he'd invited two girls from his school around and the three of them had been drinking alcohol. Just a year younger than their host, the girls weren't at all used to drinking. Jack, typically, had plied them with glass after glass of wine. Now one girl was unconscious in one bathroom after vomiting her heart out.

'She's got some kinda bowel condition,' he tells me anxiously. 'She looks awful.' The other girl is still retching in another bathroom. At least he'd had the sense to tell his sister, Susie, who'd been up in her room, totally unaware of what was going on. Susie's a bit older. And a lot more sensible. She'd decided to call us – and she'd also managed to call the two girls' parents.

'Right, we're going to get the girls out of here and get them to hospital,' I tell the siblings as my colleague, Stan, heads upstairs to find them.

'You'd better ring your parents now at the restaurant and tell them what's happening. They'll need to come straight home. We don't want to have to worry about leaving you two in the house once we leave.' Jack looks at Susie, then he looks at me. Am I mistaken, or is his voice shaking?

'I don't wanna call them. Dad'll kill me when he finds out, he'll go mental.' Susie, meanwhile, says nothing, looks at the floor. Part of her knows there's no escaping the situation. But the other part wants to help her brother, who is clearly the instigator – and the one that will get it in the neck.

'Look, parents do usually get annoyed by things like this, but they won't stay angry for ever. In the end they'll be OK,' I tell them.

'You don't know my dad,' says Jack flatly. 'He'll kill me.'

'Susie, can you ring them, please?' I say to his sister. Crews get this all the time, especially with youngsters who've done stupid things like this under the parental roof. This is an expensive house. It also looks like it's been interior-decorated to within an inch of its life. The more money there is, the greater the noise of the parents going ballistic. I do hope this isn't one of those times where the parents are more concerned about the state of their home than finding out what's really going on with their kids.

As I go upstairs I can hear Susie on the mobile, talking to her mum. They're on their way. Good. Stan, meanwhile, has already checked out the unconscious girl, Katie, and we're about to move her downstairs when her mother arrives. The woman is quite calm, despite her daughter's state. I explain the situation briefly, reassure her that Katie will be fine once we get her into A&E, and she nods.

'I thought she was at Gemma's house,' she says. Gemma, the other girl, is shaky and weak but she's now managing to walk gingerly downstairs. It turns out Gemma also lied and said she was at Katie's house.

Just as Stan and I are lifting the unconscious Katie on to the carrychair, a car pulls up in the drive. Doors slam. Then the loud, angry, booming voice of the dad in the hallway downstairs. Lots of shouting. Lots of swearing. Then, briefly, whack, the sound of a hand striking flesh. Then we hear Susie crying, defending herself.

'Dad, I didn't know what they were doing, honestly,' she pleads. Then another slap.

'This is terrible,' I say to Stan. 'No wonder that kid was scared. I'm going down!'

It's quite ugly. A huge, red-faced, clearly drunken man – driving too, his wife probably too scared to stop him – a sobbing Susie, nursing her face and the wife, attempting to comfort her daughter. Jack is standing there mute – clearly scared out of his mind. The wife, too, says nothing directly to the red-faced husband. Everyone's frightened of this brute. But it's completely understandable. He emanates unexploded violence. I feel frightened of him myself. I try to bring things down as best I can but he's still ranting and raving, swearing at his children, calling them every name under the sun.

'We can talk about all this afterwards,' I say, 'but we need to get these two ill children out of here.' This does briefly have some sort of calming effect.

'I'm sorry,' he says without much grace, then bounds upstairs to Katie and her mother in the bathroom. But one look at Katie, pale and unresponsive, and he's off again, shouting obscenities at his kids as he charges downstairs again to launch yet another verbal tirade. The children sob their hearts out.

Stan and I try to ignore the mayhem. We sit Katie down in the chair, wrapped in a blanket, and get both girls and the mum into the ambulance, lying Katie on her side in case she vomits again. Swiftly we do our checks – pulses, blood pressure, blood-sugar level – as these can become dangerously low in children if they drink too much alcohol. But so far, they're OK. We could drive off now. But I'm really worried about the red-faced man and those kids.

'I think I've left something behind,' I tell Stan. He knows it's a fib. But he knows me well enough to know that I'm really worried about what we've witnessed. So I knock on the front door again. I just need to see they're OK, to reassure us that things have settled down a bit. On the doorstep, I can still hear arguing and raised voices inside. No one answers. I try again, banging louder, and after a minute or so the mother opens the door. She's crying and gestures me to the lounge, where Jack and the dad are standing.

Jack has a large, red hand-print across one cheek. And he's wet himself. His tracksuit trousers are soaked with urine at the front. It's most likely he's wet himself through fear – there's no doubt that he was terrified of his dad's wrath right from the word go. To the trained mind, it could mean that his father had hit him so hard that Jack had lost consciousness briefly, leading to loss of control of his bladder. But I don't want to embarrass the kid, so I say nothing. The dad leaves the room.

'I think Katie may have left a cardigan somewhere,' I say vaguely, pretending to look around, 'but it might be an idea if Jack comes in the ambulance to the hospital with us. He's been drinking too, so he should be checked.' The truth is, I'm worried that Jack might have a head injury. I think it's a great idea for him to go somewhere safe – away from this ghastly man. But Jack, ever loyal, won't go.

'I'm not leaving Susie and Mum alone with him,' he mutters. His mum thinks hospital's a good idea too. She tries to encourage him to join us but he won't have it. So I have to leave and get back in the ambulance. Stan revs up and we race off into the night. On the way, I get on the radio and ask control to send police round to the house immediately to deal with an assault on teenagers by their father. And once we get them into A&E, the two girls make a fairly swift recovery from their alcoholic ordeal.

But later that night the police ring me on my mobile. I think they're calling for some sort of statement – that's the usual form in a situation where we report something like this – but this time it's different. When police got to the house Jack strenuously denied that any assault or violence had taken place. And the family all agreed with Jack's version of events. It didn't happen. I explain to the officer that this isn't true, that we'd heard the man slapping his daughter and seen the red mark on Jack's face that only his father could have made. But the police have to work by the book. If Jack and the family won't admit to the father's violence, there's nothing they can do. Sorry.

I don't give up. I then make a referral to social services about the incident. But when I contact them a few days later they say there's no case to answer. They're unable to take any further action. Sorry again.

Sometimes I drive by that road and I remember the family, outwardly so materially prosperous, inwardly living with fear of that man. As I always say, it's a huge mistake to think that domestic violence or child abuse somehow co-exists only with poverty or deprivation. It lives everywhere, yes, even in those beautiful winding, leafy roads with the huge, detached houses, the security gates and the tastefully furnished rooms.

A RANDOM ATTACK

The little boy has the face of an angel: enormous brown eyes and long, curly eyelashes. He's a sweetie, casually dressed and quite calm. Despite the fact that a knife-wielding maniac has just slashed him across the face for no good reason.

It's 11.30pm. We've been called to a parade of shops where a ten-year-old boy has been assaulted. We know he's got a facial injury. But it's all a bit strange. What was a ten-year-old doing in a public place at this time of night in the middle of the week? Hasn't he got school the next day?

The police get to the boy first. They've even managed to put a dressing on the lad's cheek. He's covered in blood, lying there on the pavement with the officer holding the dressing.

'This is his brother,' the copper says, gesturing to an older boy, around fifteen. He had gone after the lunatic who carried out the attack. Luckily he didn't catch him. The police fill me in. The two black kids were in the street, outside the shops, when the stranger, a young white man in his twenties, had come up to them, started yelling and screaming and waving a knife around – then he'd lunged at the boy with the knife, slashing him on the face.

Meanwhile, a bystander had called 999. Now a small group of people are clustered around us, staring. The little boy is so quiet, it's almost disarming. The knife has gone deep into his cheek, all the way through, leaving a large, gaping wound. It's not life-threatening. But he'll probably need plastic surgery to repair it and minimise any scarring. The wound does look really nasty now, but once the doctors have closed it he will look better, and he's young, so it will heal well.

'We went out to buy crisps,' he tells me in the ambulance as his older brother comes to sit with us. The older boy, however, feels responsible for this.

'I shoulda caught him,' he tells me.

'Thank heavens you didn't – he could have killed you,' I say, feeling strangely upset that these two young kids had found themselves under such a random attack – for what? Nothing, by the sound of it. The older boy has already given police their mother's number and we wait in the ambulance for her. Normally we get a distressed, nervous parent turning up, distraught at the news that their child has come to any kind of harm, wanting to cuddle or comfort. But life is never quite what you expect, is it? Here's the mother, a large maternal-looking woman in her forties. Before I know it, she's ripped the dressing straight off his face.

'Oh what have they done!' she says, throwing her arms around him, falling to her knees. The poor little lad is seat-belted to the trolleybed. Yet he doesn't react at all. No tears, nothing. It's all very odd.

This is a bit weird, I think, as I try to unpeel her from him.

'Let go! You're pulling on the wound,' I plead, attempting to replace the dressing. She ignores me. Her attention is now focused on the older boy. Then, without warning, she starts to lay into him.

'You let this happen, didn't you?' she yells, punching the boy around the head again and again.

'Why did you let this happen, eh? Why did you let them do it?'

The kid is trying to shield himself. 'Mum, don't! Mum, don't! I'm sorry, Mum, I'm sorry.' But she's furious, relentless, she's intent on whacking him again and again. This is truly awful. It's already out of control. I manage to pull her backwards off him, wrap my arms around her.

'Stop hitting him! You mustn't hit him!'

'He's my son!'

'Yes, but you can't hit him,' I reason.

'Let me go!'

'I will if you stop hitting him.' Reluctantly I let her go – only for her to start laying into him again. And he's sobbing, trying to move away from her blows. It's just terrible. A colleague leaps out of the driving seat to help me. And we just about manage to drag her off him and open the ambulance doors to get her out. We then tell the police officers outside what's gone on. And we make it clear she can't come in the ambulance to the hospital. She's continuing to yell and carry on, though the police do manage to quieten her a bit.

Yet in the ambulance the little boy is still very quiet, overwhelmed by what's happening. The older boy, however, is quite distressed now and continues to sob. His mum has blamed him for all this – and now he really does feel guilty.

'I shoulda gone after him,' he sobs again and again as we finally set off for the hospital. Once there, the two boys are taken to a cubicle to wait for treatment.

Later, back at the hospital with another patient, we ask after the boys. 'Oh, they're fine,' a nurse tells me. 'The mother never turned up. But someone, think it was an aunt, came to see them.'

Afterwards we learn that the mother has a history of mental health problems – which does explain some of it. And the knife man is caught, jumping on a bus. A driver spots the knife and calls the police. So he doesn't do any more damage.

There's no bad ending, yet it's not, for me, a run-of-the-mill job. You hear a lot about all the dreadful things that happen to kids in London but the truth is, this is the first time I've found an innocent ten-year-old attacked on the street like that late at night. I felt bad for that little lad. How scary it must have been for him, first to have his face cut like that, then to be in hospital without his mum there to give him a loving cuddle. It's scary enough for a kid to be in the emergency department at midnight without having watched his brother be attacked by his mum.

In the end, you just hope those kids are resilient. If they're fairly robust they might be OK in life. But you do wonder.

AN ALBUM

Quite often people ask me, 'What's the worst job you've ever done?' Surprisingly perhaps, it isn't the jobs with blood everywhere or terrible injuries that upset me the most – although working with victims of violence who are in terrible pain or dying can be awful. And, of course, seeing children in extreme distress is equally troubling. But sometimes, being confronted with emotional pain is just as bad, if not worse, than dealing with the effects of physical injury or pain. When it's physical, a paramedic can usually do certain things on the way to hospital even when we know it's pretty hopeless. But when it's emotional there's nothing in the world you can do.

The middle-aged couple are terribly apologetic. They hope we don't think they're being nosy parkers. But they thought it best to ring 999. 'We didn't know what was wrong with her,' the woman tells me. 'We were worried that she'd had a heart attack or something.'

Richard and I have been called out to a busy shopping

centre on a warm summer evening. The couple, late-night shoppers, have noticed the lady lying prone on a bench as they make their way to a department store. An hour later they leave the store – and realise that she's still lying there. She hasn't moved. The shops are closing and the crowds are making their way home. They try to talk to her, ask if she's OK. But though her eyes are open, she doesn't answer them.

We see the woman, lying on her side, staring into space. She's about forty, wearing a light, smart jacket over jeans and a t-shirt. She doesn't look like the average bag lady. Kneeling down next to her, I introduce myself. No response. The smell of fabric softener wafts up at me from her clothes. Her hair is scented, newly shampooed. She's definitely not homeless or a vagrant. And there's no sign of any alcohol on her breath. So why is she lying down here, silent, in this public place? It doesn't really add up. Medically we're going through a tick list. A stroke? Or diabetes? The second can cause confusion if the blood-sugar level plummets dangerously low.

Richard tries to coax a few words from her too. Nothing. I tell her I'd like to test her blood sugar and keep trying to get her to respond or say something. But all to no avail. Yet she doesn't reject me or fling me away when I do the simple test. She just lies there, mute and unresisting. Her blood sugar is normal. The pupils of her eyes react normally. She seems to be following us with her eyes, however. And once or twice I see a tear fall.

'Are you in any pain?' says Richard gently. Silence. We're at a bit of a dead end. If she can't or won't tell us anything, it's difficult to work out the next move. We decide to sit her up. Obligingly she makes no resistance.

'Do you think you can stand?' I ask. Amazingly, she stands up wordlessly and lets us walk her outside the shopping centre to the ambulance. At least we can look at her in the privacy of the vehicle. On the way out we thank the good Samaritans who have been politely watching us from a distance – they seem genuinely concerned, not your average gawpers.

We lie the woman down, carry out all the usual tests. All normal. No clues. I'm starting to wonder if this isn't some kind of mental illness.

'I need to look through your pockets, if you don't mind?' I ask her. When a patient isn't in a position to help us themselves, we have to go through people's pockets for some form of ID or medical details. Strangely she carries no bag with money or keys with her. Have they been stolen from her as she lay on the bench? But in the pocket of her jacket I find something. It's a small photo wallet, with plastic-covered pages. Maybe there'll be some clues here. First, a picture of the woman, much younger, proudly holding a tiny newborn baby.

'Is this you here?' I ask, hoping to draw her out a bit. No response. I carry on flicking through the wallet. There are lots of photos, classic family shots, mostly of the child, a little boy, as he's growing up. First, a toddler surrounded by Christmas presents, then, on holiday in a paddling pool, a bit older. Then a sunny day in the park. He's wearing football kit, one foot atop of a ball. Then come the bog-standard school photos with dodgy haircuts. Then there's one of the woman, sitting on a sofa with her arm around the blond boy. He looks about twelve, the same age as one of my boys.

But then the photos change. The boy is still smiling up at me, but he's in what looks like a hospital bed. More photos in hospital. Then a photo of him without his lovely mop of thick, blond hair: the effects of chemotherapy? I wonder. Oh no. Does he have cancer, leukaemia maybe? Turning the pages is like going through a story book without words. Tragedy unfolds with each page. He's lost weight. His condition is deteriorating. One of the photos shows him hooked up to a drip. He's smiling. But he's a very sick lad. I estimate he's about fourteen now. Finally the last photo: he's not smiling any more. His sad face just stares at the photographer, gaunt, thin, a shadow of the happy boy in the park, beaming so proudly in his football kit.

Only now do I understand the woman's silence, her unwillingness to communicate, her withdrawal from the everyday world. It's grief, profound, unfathomable, numbing loss – the loss of a beloved child. As a parent you can only start to imagine the pain and suffering she feels at the death of her beautiful and precious son, the only outward sign manifest in the occasional tear falling on to her cheek. And my guess is, it's happened earlier today.

I want to be compassionate, to say something. But for once I'm lost for words. Any words I can use would be too trivial, too patronising in the face of her deep and terrible sorrow.

Paramedics are used to saying certain things at certain times in an attempt to calm, to soothe – or to warn relatives or partners. But this time there are no words that will help a grieving mother who has watched, week by week, month by month, as her adored child fades away before her. I put the album back. All I can do is hold her hand as Richard drives us to hospital.

There I book her in and say farewell. She still doesn't say a thing. I reckon there is no pill, potion, test or procedure for what's gone wrong with this woman. Only, if she's willing, a referral to an understanding grief counsellor.

There aren't that many jobs that make me want to cry. But I walk out to the ambulance lost in thought. As I do so, someone comes up to chat. It's a man I know, a long-serving paramedic with a reputation for being unblinkingly tough in the face of the nastiest of call-outs: the last person you want to run into when you're feeling weepy about a call. But seeing him, I burst into tears.

'What's wrong, love?' he says. I apologise and briefly tell him the sad story of the lady and her photos, half-expecting him to mock me for being such a softie.

He doesn't. He's amazingly kind. He understands. 'We all have a particular job that gets to us,' he says wisely. 'Even me!'

'You! I can't see you getting upset about anything!' I tease him.

'No,' he assures me, he too has his moments. 'But if I tell you, Lysa, I'll have to kill you!' Of course. The seen-it-all image, the protective shell most paramedics don, along with our uniforms, our stab vests – our armour if you like – must be maintained at all costs. I know where he's coming from. You couldn't do this job for long if you gave into your own feelings all the time.

Then he offers to get me a cup of tea, the perfect gesture. It may be a Great British Cliché. But there's something about a cup of tea that can make most things just that little bit better.

ON THE PLATFORM

A&E, New Year's Eve, is a bizarre experience for those who work there. You spend the first part of the night with virtually nothing to do, knowing full well that this is the calm before the storm. Then, half an hour after midnight, World War III starts with a vengeance as the revellers pour in. It's absolute mayhem, a deluge of drunken, incoherent or verbally abusive people, most of them covered in blood, vomit or their own urine. Some are half-standing, some are on stretchers, some are accompanied by noisy, drunken friends or abusive relatives. There's a lot of shouting and swearing from everyone.

Tonight half a million people are expected to pour into Central London to see in the New Year. This year I'm working in a treatment centre we've set up in Waterloo Station, in a cordoned-off area, so we can deal with the drunks, the assaults and the general chaos. We've started doing it this way recently because the hospitals are overwhelmed by revellers. Having us in various hot spots around the centre means we can treat people on the spot and keep some of them out of A&E.

The night is still young. Around 11.30pm I'm asked to go and help a woman on one of the station platforms. She's having a fit apparently. With another paramedic, Carole, and a St John Ambulance guy, I make my way through the station throng. It's already heaving, worse than rush hour or a big football match. We battle our way through till we find her, sitting on a bench. She's got two young children with her, a boy, four, and a six-year-old girl.

The mother is a skinny, slight woman in her thirties, very long, fair hair, all glammed up for the night: tight, party sheath, furry bolero, definitely out to party. The little girl is dressed up too, in spangly frock and disturbingly high heels.

'Are they your mum's shoes?' I say in amazement.

'No, they're mine.'

'Can you walk in them?' I ask, wondering how many six-year-olds are trotting around tonight in high heels.

'Yes, but my feet hurt.' She shows me her heels. They're bleeding. But there's no time for a plaster because Mum is crying, sobbing her heart out. She is also seriously drunk. The smell of booze is oozing from every pore. The party obviously started a bit too early. I get an idea of the story from the little girl. Mum's not really making sense.

They'd caught the train from Kent to come up to town to see the fireworks.

'And Mum had a fit when we got here. She's had them before.' Hmm. I'm not too sure about this. I think she's just pissed and throwing in a bit of drama. Sure enough, Mum starts to have another 'fit' in front of me, throwing her arms around and kicking her legs out. It's not very convincing.

'I think we need to get you some help,' I tell the woman.

'I don't bloody need help, leave me alone – and don't touch my kids!' she yells. Then she promptly launches into another fake fit. I've seen hundreds of real fits in my time. This second attempt to convince me with more arm waving and kicking out doesn't work, I'm afraid. Mind you, fit-faking isn't at all unusual. I've even seen people pretend to be unconscious. People start doing it for effect or attention. Then, once the paramedic or ambulance arrives, they're backed into a corner. So they have to maintain the fiction. Does it happen often? 'Fraid so.

But we can't just say, 'Oh, she's faking.' We are here to help. Somehow Carole and I get struggling, swearing Party Girl on to a trolleybed and cover her with a blanket, even though she's insisting she doesn't want our help.

'We're fine, go away, we don't need you,' she hisses at us. The kids, meanwhile, are quiet. Holding on to the trolleybed as they try to keep up, we all manage to get through the crowded station. The plan is, the St John Ambulance guy will then take over and get her to hospital. It's all going well until Carole and I actually get them into the ambulance. We seatbelt the children and I'm just about to warn the St John guy that when they do get to hospital, child protection concerns should be officially documented. OK, they'll probably think of it anyway.

'But someone in this state in charge of two very young kids could be a problem,' I start to say. Then, whoosh, it's all gone off. Now the woman's going berserk, flailing around, arms and legs waving, screaming her head off, eyes screwed shut. She manages to kick her small son in the chin in the process, though he barely flinches and continues to sit there, quiet, resolute. The little girl, however, is trying to quieten her mother.

'Mum, Mum, no, please don't,' she pleads, painful for any onlooker to hear, a small child begging a parent to please stop the nightmare. And, racing through my head, is the thought, How often have these children already lived through scenes like this?

'Don't you dare take my kids away!' screeches Party Girl. My instinct was right. This isn't a one-off. It's all happened before and she probably realises – anyone found drunk in charge of minors is likely to be arrested. First and foremost, there's the safety of the kids to consider.

As I bend down to ask the little boy if he's OK after the blow, she lands a powerful kick, whack!, right on the back of my head. Luckily she's wearing ballet pumps and not stilettos. But I fall over on to my knees, trying to reason with her as I come up. Carole tries to help me.

'Look, your kids are here, they're coming with you. That's not what we do, we don't take children away,' Carole tells her in an attempt to stop the flow of insults coming from her. But no, she carries on screaming, oddly with eyes still tightly shut. Then, wham! Without warning her leg comes up and kicks me right in the stomach. It winds me temporarily.

Oh, no, where are the police? I'm thinking now, knowing that hordes of police are inside the station anyway, as the little girl carries on begging her to stop.

'Mum, don't kick the lady… Please, Mum, don't.'

Her distress is so painful, so poignant, the attack on me seems less important – until Party Girl kicks me again as I turn to try and comfort her young son, this time between the shoulder blades. She manages to kick Carole too. Things are getting out of control. But the police can hear the mayhem from outside the ambulance. One puts his head round the door.

'Er – shall I book her now, or do you wanna go to hospital?'

'Well, she's apparently had two fits, so it has to be hospital,' I warn him. But I'm too shaken up to go with them. Carole says she'll take over and bizarrely the woman stops shouting and calms down. Maybe the fact that I'm off the scene quells her fears. Later Carole tells me that Party Girl hugged her tight all the way in. Bizarre.

Afterwards we learn that when she got to hospital, she attacked three members of hospital staff and police arrested her. And the kids had to be taken to a place of safety until staff could organise foster care. Maybe a relative was asked to come and collect them, but we didn't know any more than that. All very sad: a New Year's treat to see the fireworks ending with Mum in a police cell and the kids possibly taken into care.

The next morning when I wake up I can't move my neck. It's really stiff. Is it whiplash from the kick? Or something worse? After some painkillers I deduce it's whiplash. No real harm done. At the end of the day, what happened is an occupational hazard for any paramedic – particularly on New Year's Eve. Most paramedics have been hit at some point. And verbal abuse and threats are really common. But in all my time in the job I've only been hit a handful of times – that was the worst. People who work regularly on weekends cop a lot more violence than me.

But I'm still thinking of those kids. Obviously they were used to this type of thing. The little boy was so subdued through it all, not a tear, not a whimper. And the little girl was more like a teenager – a six-year-old desperately trying to parent a naughty mum. As for those fits, who knows? I'm pretty sure that what we saw was just bad acting, something a person switches on or off, mainly for effect. In a way, that's the saddest thing of all, pretending something bad is happening to you because you can't see any other way out of the rotten situation you've got yourself in.

CLOSE TO THE EDGE

THE AXEMAN COMETH

Inner-city ambulance and A&E staff are right on the frontline alongside the police when it comes to dealing with dangerous, violent people in the course of the job. It doesn't happen every day of your working life – some paramedics I know face a lot, others seem to work for years without too much of it – but it is something we are trained to handle. Even so, there are times when even the most dangerous, violent individuals don't present themselves to us that way. And that means that we don't pick up the danger signs – until it's almost too late.

Here's a crazy call, an attempted hanging in a pedestrianised shopping mall. What a public place to try it! But this is what comes through tonight from the police. He is a balding man in his sixties, sitting on a wall. He is extremely tall, six foot four or more. He attempted to hang himself from a nearby lamp-post. But he didn't really think it through. He used a trouser brace round his neck and attached it to the lamppost. Of course, it didn't work. He was too tall and the brace was made of elastic. And it was an incredibly busy place. So passers-by intervened, kept him talking and phoned 999. Now he's calmly having a fag, talking to police, and they're filling us in on the story.

'No,' he says firmly. 'No hospital, thank you very much.' He's really sorry he has wasted everyone's time. But he wants to go home now – and go to bed.

'Look, you may have injured your neck or something. It's best to get a doctor to have a look at you,' I cajole. He looks at me quizzically, as if it hasn't occurred to him that trying to hang yourself in a shopping precinct comes under the heading of dangerous sport.

'OK, I'll go,' he says with obvious reluctance. It's not quite as simple as that, however. Attempting to hang yourself is viewed by the authorities as a mental health issue. It's not exactly normal. So the police are obliged to offer to come with us in the ambulance, as a safety precaution. But tonight I'm in Barbie medic mode. The man appears quite harmless, although he is a well-built muscular man, fantastic posture. He may have his problems but I don't sense any malicious intent. Sure, he looks like a baddie with the bald head and his height, a bit like Blofeld in the Bond films, the character that wears a monocle and strokes a cat. But he's not unpleasant in any way.

'No, it's OK,' I tell the police officer. 'We don't need them, do we?' I say to the man with a smile. He just nods.

But on the way in, out of the blue, he surprises me with a question: 'Can I hold your hand, miss?'

I reach out, take his hand. 'What's the problem?'

'I'm scared,' he says, all six foot four of him. Then there's a long pause. 'Er… I don't think they're gonna take me at the hospital.'

'Why's that?' I ask in my innocence.

'Well, I'm barred from that hospital.' He's talking nonsense. I'm beginning to think he's a bit of a pussycat, to tell the truth. I tell him I can't believe that. He's a bit daft – but totally harmless. 'Last time I was there they told me to never come back again.'

'Don't worry, you're with me this time,' I coo reassuringly. 'And anyway, why would you be barred?'

'Well… I tried to kill a doctor last time.'

I'm still not buying into all this. But it's best to humour him. So I go along with it. 'How'd you do that, then?'

'I tried to cut his head off with an axe.'

And no, with all the training – and all the unstable people I've dealt with – I'm still not alarmed or worried. People say the strangest things to you all the time in this job. And anyway, I reason, if that was true, he wouldn't be sitting here with me now, would he?

Now we've swung into the hospital. Once outside, I link my arm with his for extra reassurance and we walk into A&E together. It's as if I'd walked in arm in arm with Saddam Hussein. Every single head working in the department, half a dozen staff, swivels round, looks at us – and promptly dives for cover. People vanish around corners. Doors close. Is this a joke? Here we are, standing in a nigh-deserted A&E reception. Just me and my tame axeman. Who, I'm still convinced, wouldn't hurt a fly. Then a head pops out of a door, a nurse I know by sight.

'Get him out of here now! I'm calling the police! Just get him out! Take him to another hospital, we're not having him here!'

'But we can't, we're here now,' I protest stupidly. The man says nothing. He just looks at me as if to say, 'Told you!' Our arrival must have triggered alarm bells, however, because hospital security guards are restraining the man before the police arrive. They immediately grab the man and cuff him. I'm standing there totally bewildered. The staff reassume their positions. And then one of the nurses comes over to me.

'You know you really shouldn't have brought him here, Lysa,' she admonishes me.

'What am I supposed to do with a man who's just tried to hang himself with a possible neck injury – leave him in the road?' I say, starting to get angry.

'You obviously don't know what happened here last week.' Then the penny drops. I've been away for a few days so of course I wouldn't have heard anything on our grapevine. The reality is that my Mr Harmless is in fact Frankenstein's monster.

Somehow he'd managed to walk into a busy A&E the week before, clutching an axe. Then he'd gone for one of the doctors and started to threaten him with it. Bravely one of the nurses had intervened, managed to grab the axe mid-swing. Police had arrested him, but then he'd been released. And the staff still haven't had time to get over the shock of it all when in marches Lysa with the man, all smiles!

When I get home that night I'm pretty shaken. How could I have missed it? I just took him at face value. Could have been a big mistake. I recount the story to my husband.

'You're too trusting,' Steve tells me. 'The police were there. Surely you could have put two and two together?' I didn't. And why they let him back into the community so quickly after the A&E incident is beyond me. The thing is, he's still living in the area. To this day crews and A&E staff still talk about him. Other ambulance people sometimes tell me they've been called to his home now and again.

The doctors say he's got some sort of personality disorder rather than a mental illness. So he can't be treated by medication. A personality disorder is all about learned behaviour, rather than a chemical imbalance in the brain, which can be treated by drugs. Paramedics who've been inside the house say it's a bit of a shrine to the army, with heaps of memorabilia on display. And I'm told he likes to walk around in a military uniform.

So there you have it: a mad axeman who can plausibly present himself as Mr Harmless. It makes you wonder. But it sounds like someone in the medical profession believes he's harmless too. They'd better be right.

SUPERMARKET DRAMA

Christmas in a few days and here comes a call for anyone in the area to contact the control room. We're in the middle of our paperwork, then we get the job: all we know is it's a very busy supermarket in Croydon and a man aged 59 is unconscious. That's it.

'It's a cardiac arrest, I know it,' I tell Dave, who's been working with me all day. The traffic's bad, we're busy, law of averages says it's someone in cardiac arrest. Halfway there and there's a bit more info: a rapid-response ambulance has now arrived at the supermarket – and, yes, the man is in cardiac arrest. For paramedics, this situation can be fairly complicated. We may have to pass a tube down the man's throat, inject certain drugs – or even give him electric shocks. So Dave and I talk through what we think we might need to do, just to be sure. Outside the supermarket we're caught up in a frustrating sea of shoppers unloading packed trolleys, and it's a battle to get the vehicle close to the door, what with all the equipment we'll need to bring in.

But once we're in, staff have actually managed to screen off a little area close to the doors. And a St John Ambulance person has also got involved. The guy was actually buying himself a sandwich in the supermarket when someone stopped him and asked him to help. The cardiac arrest case is on the floor, unconscious and flat on his back, surrounded by paramedics and supermarket staff. He was going around with his trolley, had a sudden cardiac arrest, banged his head on a railing as he went down – and wound up lying there, not breathing and without a pulse. The staff called 999 and, as directed by our control room, one of the women from the shop floor managed to do resus, compressing the man's chest.

Then the St John Ambulance man was discovered and began CPR, using his own equipment, bag and mask, to try to push oxygen into the man's lungs. Steve, our rapid-response guy, turned up within minutes too. He quickly put two sticky pads on to the man's chest, plugged in his monitor – and discovered that the man's heart was fluttering like a bag of worms – and not pumping blood into the body and oxygen into his brain. The only life-saving option now is an electric shock.

Steve then tells everyone to stand clear and gives the man's heart an electric shock (called defibrillation) through the pads. One shock, a thud, then nothing. Try again. Thud: this time, success! The shock jerks the man's heart back into a normal rhythm. And it's beating for itself again. Technically this man was dead – if salvageable. With a stroke of blinding luck, the right person has turned up and intervened in the nick of time. But it's still a knife-edge situation: he remains unconscious and very poorly.

Dave and I are being filled in on exactly what has happened and, though there's still a huge risk that the man's heart will start fluttering again, we can take over. Right. A cannula into his arm – that's a route for me in case I need to give him any cardiac drugs or fluids. A good job Dave and I had talked it all through as we drove in. There are five of us, but everyone in the team has to be on the ball about what they're doing. Time is against us. This man could easily have another cardiac arrest. He desperately needs the resources of a hospital resuscitation room.

I want the St John guy to come with us because I'm driving and Dave and I will need his help if the man does have another cardiac arrest. The man is bleeding too from the back of his head and needs dressing and maybe stitches. But that's a side issue – the main thing is to get him to resus really fast.

Luck seems to be on everyone's side tonight. Just as we're getting ready to leave, the man's wife turns up. The supermarket staff had cannily managed to get some phone details from the man's loyalty card. All things considered, she's pretty calm. But she's a bit confused because he'd gone out planning to go to a different supermarket. She can't work out why he's here.

'Maybe they didn't have the chicken,' she keeps saying. I try to explain it all to her, that he's not out of the woods yet. I also explain that the blue lights and sirens will be on all the way as she joins us in the ride to hospital. People can get very upset about all that, so sometimes it's better to warn them. Then a shopper comes over and offers to help the woman and to park her car safely. It's a decent thing to do for a total stranger and the bewildered wife is grateful for the help. Who says the Christmas spirit doesn't exist? I breathe to myself.

And yes, we do get the man there, heart still beating, no second cardiac arrest, straight into the resuscitation room to recover. A scan afterwards shows there's no brain damage. He'll pull through. It's just an hour's work out of my day. And not that unusual a job for a paramedic. But for the couple it's a small miracle. For them – and their family – this Christmas will be especially poignant. Thanks to the quick-thinking response of strangers, he got CPR almost from the moment he dropped. For the truth is, that's when people survive cardiac arrests like that, when bystanders know what to do. And then an emergency person turned up to get the electric shock to the heart really quickly. Even if all the staff did was push up and down on the man's chest until the ambulance arrived, that's better than nothing. And 999 control room staff always stay on the line to give callers instructions on how to do it.

I keep thinking about it afterwards, as you often find yourself doing after jobs where every minute matters. OK, the man might still have survived without that early CPR – but probably he'd have been left with brain damage. We might have managed to get a pulse back. Just. But six minutes without oxygen is six minutes too long. So while the shock did the trick getting him back, no question, the supermarket workers were the ones who managed to keep him ticking over. They were the real heroes of the hour. One Christmas shopping afternoon they'll all be remembering for a long time.

THE NEEDLE

Our work has evolved over time. Traditionally ambulance people went in with a trolley and put someone on it, and that was the job. Steve, my husband, says that in the early days all you needed was a driving licence and a first-aid certificate. But those days are long gone. Nowadays UK ambulance teams are highly trained and skilled in various treatments, medications and life-saving interventions which we administer before getting people to hospital. As medicine and technology have advanced, so have our skills. Sometimes you find you're trained to do something specific – but you may only get to do it once, maybe twice, in hundreds of emergency call-outs.

I'm driving around, alone in the car, mid-summer when I get this call: 'Thirty-eight-year-old man, chest pain.' It could mean either a cardiac or lung problem.

Standing waiting for me in the front garden there's an older, white-haired woman and a boy, about eight or nine, looking anxiously at his dad, who sits on the doorstep. Oddly there's no front door. Then I see it propped up against a wall.

This man, very tall, t-shirt and jeans and lots of tattoos on his arms, looks quite ghastly, pale, sweating and obviously quite ill. He had a sudden pain in his chest – on the left.

'Me and my boy were hanging the front door for the old lady when it started,' he explains. He'd also had a chest infection for a couple of weeks and taken antibiotics. Hmm. Could this be cardiac-related? Or just the chest infection? I can't discount a clot on the lung either. I check his chest and it sounds normal. But there's nothing normal about the way he's half-hunched over with pain. Also, his breathing is shallow, so I get an oxygen mask on. It seems to help.

Then my back-up arrives, an ambulance with a paramedic and a technician. We get the man inside the vehicle and are just getting ready to start up when everything suddenly changes. Now his breathing is much, much worse. Between gasps he's telling me he's in excruciating pain now, in fact he can hardly talk for the pain and the effort of breathing. It's getting really serious. Quickly I cut off his t-shirt to have a proper look at his chest: this could be a collapsed lung. At the same time the white-haired lady looks in – and sees me cutting her son's clothes.

'Oh no, what are you doing?' she says, clearly alarmed and shocked. The boy, trying to be useful, is running in and out of the house, tidying up Dad's tools. I murmur a few reassuring words and the mother retreats.

I close the ambulance door as I need a proper a look at his chest fast. If both sides of his chest are symmetrical, it's not likely to be a collapsed lung. The problem is he's covered in tattoos which are making it difficult to assess the symmetry of his chest. So I inch myself behind him, looking over his shoulder and yes, the left side of his chest is more expanded. It's called hyper-expansion. And it's bad news. It means that the lung on that side has collapsed. The air is still going into the lung but once it's in there, it's escaping through the hole in the lung and getting trapped between the collapsing lung and the chest wall. It's like air entering a balloon inflated inside a jam jar; if the balloon has a hole in it the air would get trapped between the balloon and the jar and be unable to escape. And it's getting worse with every breath he takes, because more air is going in than can be breathed out. He's in terrible pain. And he hasn't got long.

'I'm going to have to relieve the pressure,' I tell him. Three pairs of eyes are looking at me. My colleagues know what I mean. But the man doesn't. And he's too far gone to start asking questions. I have to do this fast.

Passing a needle into someone's chest to relieve pressure isn't an everyday procedure. Some people who do my job never get to do it at all, though it's part of the training. I have done it before – just once. But the person was unconscious. This man is looking at me, an 'Am I breathing my last breath?' look; a look we never want to see. His breathing now is incredibly shallow: it's too painful to take a deep breath. And his lungs have nowhere to move because of the trapped air.

'I have to do it now,' I explain. 'There's not enough time to get you to hospital. I have to pass a needle into your chest. It'll relieve the pressure straight away.'

'Just do it,' is all he says.

The needle I use is the largest we carry, about three inches long. It will hurt as it goes in but it's nothing compared with the pain he's suffering.

'Look at the ceiling,' I tell him, outwardly confident but inwardly thinking, Lysa, you can't afford to stuff this up. Then I insert the needle between his ribs, through the chest wall, into the air space. Immediately there's a reassuring hiss as the air escapes through the needle from the chest outward. The hiss tells me I did not misjudge the situation. It is the right diagnosis. Now, with every breath he takes, he's starting to feel a little bit easier as the pressure in his chest is released by the needle. I check: yes, his respiratory rate is coming down too, a good sign.

It's an understatement to say I'm relieved. My colleagues look at me. I can see their relief too.

The man's pain is decreasing. But carrying out this procedure is a time-buying exercise. What he needs as soon as possible – within minutes – is a chest drain inserted in hospital, a much wider conduit for air to escape through. I open the door to have a quick word with his mum outside. She's now sobbing her heart out. She hasn't been privy to the small drama inside and has been imagining the worst.

'Is he… dead?' she says, her eyes pleading with me.

'No, he's a bit better, but we've got to go to hospital now,' I say.

'Oh, I can't lose another child! I can't bear it!' she sobs. I don't know what she's talking about. But there's no time left to find out. And she can't come in the ambulance – because there's no front door to the house, she has to stay behind and look after the boy. So we rev up and rush off. A doctor is standing at the hospital doors, waiting for us. I brief him quickly as the others lift the man out. But the doctor isn't happy.

'Why the needle?' he says sharply. Most simple collapsed lungs get better on their own.

'The needle helped,' I say firmly, though I do begin to doubt my descision. But I know it wasn't a simple callapsed lung. There was tension and his enlarged lung was putting pressure on his heart.

This job's over now and I go off for coffee, a bit perplexed by his attitude, but still convinced the needle was the only option. But when I get back after my break, the man has taken a turn for the worse. They had to put a chest drain in very quickly, there wasn't even time for a local anaesthetic. But the drain did a better job of releasing the air than the needle did. As I leave for the next job, I run into the doctor who'd queried the needle.

'You were right. There wasn't much time to play around with,' he says. Then he winks at me cheekily, just to show me what a lovable guy he really is. Because the man looked quite well when he came out of the ambulance, the doctor didn't realise just how close to death the patient had been just a few minutes earlier.

The trouble is, doctors don't always appreciate an ambulance environment or, come to that, anywhere that isn't a hospital. Their world is always well-lit, dry, clean, with help to hand and equipment ready. They don't know what it's like trying to examine someone's chest while noisy, deafening traffic rumbles by. Or how difficult it can be to carry out a life-saving procedure with 50 bystanders looking on with useful comments like, 'You want to get him to hospital, love' or, 'Do they let ladies drive ambulances, then?'

But this man goes on to make a full recovery. And a nurse told me later that his mother had indeed lived through a previous tragedy. Sadly her older son had dropped dead in that same front garden, a sudden heart attack at 45. She must have thought her worst nightmare ever was being repeated in front of her all over again.

DANGER ZONE

It's windy and dark. Andy and I have no idea who called 999 or why. We just get to the address in a small block of flats. It happens sometimes: people just ring 999, give an address and hang up. They've done their bit, don't want to be involved any more.

The front door's been left wide open. In the lounge a man sits on a sofa, hunched forward. He's a big guy, really beefy, about forty, scruffy, unshaven, unkempt. He's wearing only vest and pants. His most outstanding feature is his very hairy back. There's a coffee table right in front of him with glasses, ashtray and a couple of bottles of vodka. One's on its side and empty, the other's half full. He's stares at us but remains silent. For all we know he's been sitting here like this for hours. Tanked up. We start by trying to talk to him, usual stuff: How are you feeling? What's wrong? Are you in pain? But there's no response. He just sits there, looking at the table in front of him. It's a pathetic but nonetheless sad sight. And quite sinister too.

I perch down next to him. He reeks of booze. Maybe I can coax something out of him. Andy just stands there, a 'what now, Lysa?' look on his face. I'm just starting to wonder what we can do next when the man starts talking, rocking backwards and forwards, quite distressed – and barely coherent. He's not exactly talking to us in a conversational tone. And it's hard to make much sense of what he says because he's muttering and his English is heavily accented.

Then we catch something about 'the voices'.

'They got it in for me, they don't like me. I must hurt, kill me.' Oh dear. Fragmented as the sentences are, these are the sounds of someone who has flipped. Badly. You learn to recognise the signs: looking closer, there's a horrible look in his eyes that tells you this man is dangerous, to himself and to anyone who gets in his way. And almost simultaneously, as the penny drops that this is a potentially volatile individual with mental health problems, I see it.

It's a knife.

It's a big one, probably a bread knife. He's twiddling it in his hands as he jabbers away. If you face him, you don't see it because his hands are concealed under the table. Shit. My eyes go to Andy. I know he's seen it too.

'Don't worry, nothing can happen to you now, we're here,' I say softly, my mind racing for an instant solution. I'll try the softly-softly approach. It's our best bet.

'Why don't you put that knife down on the table and we'll be able to help you?' I say. My voice doesn't betray me. Inside I'm praying: please, please, please let this do the trick.

Incredible. It works! He places the knife on the coffee table, picks up a packet of fags and gets one out. Relief. Then he launches into another stream of gibberish, nonsense stuff, mangled words. But we're halfway out of the woods now.

Andy doesn't waste a nanosecond. He grabs a lighter lying on the table, says, 'Here's a light, mate,' and in that brief moment, as the man holds up the ciggy for the light, I manage to lean forward, cover the knife with my forearm and take it back a bit. Then I grab it.

'I'm just going outside to check if the other ambulance is coming, Andy,' I say, the knife concealed at my side. Once outside, I open the ambulance door and drop the knife between the seats: probably the police won't be wanting it as evidence later on – but you can't just chuck it out on the street, or even in a bin. Then, shakily, I grab the radio and ask for urgent police help. My voice stays calm. But I'm not. My knees are still shaking. And I'm well aware that this is not the end of it. There are two bottles lying on that table: potential weapons for a madman without a knife who might suddenly go totally ballistic.

I go back. The man is a bit calmer, doesn't seem to be bothered about the knife or where it's gone. So we get on with it. Between us, Andy and I say all the right words to convince him he should come with us to hospital. He's agreeing. We even manage to suggest he put on some clothes – there's a jacket and trousers lying on a chair outside the room – but he shrugs that off. But now, one on each side, we're walking him, barefoot and semi-dressed, out of the flat.

But as we start to walk out of the block, we see the road in front of us is chocka: police cars. They'd approached silently. I'd warned control, 'Don't use the sirens in case he panics.' And, of course, that's exactly what happens.

Spotting the cars, he flings us both away, starts shouting, 'Why police, why police?'

We can't just hand him over to the police, it'll be mayhem. Yet, for some inexplicable reason, he seems to trust me and Andy. 'It's all right, mate, they won't go near you,' says Andy and the man lets us take his arms again and edge him towards the ambulance door. The police stand back, let us do our stuff. Sure enough, he goes in quite meekly. And luckily we're only round the corner from the hospital – with the police right behind us. So we walk him into the hospital and hand him over without further ado, relieved it ended so peacefully. For us.

But while we may have defused the situation temporarily, it blows up afterwards. Later that night the guy goes completely berserk, running around the hospital car park, terrifying everyone until staff manage to tackle him. Eventually he's sectioned and taken off to a psychiatric ward. This is what's known as an acute psychotic episode. He'd previously been diagnosed with schizophrenia. He was taking regular medication for it and doing fine. Until he decided to hit the booze – the worst possible thing he could do to himself.

The trouble with medication and mental health problems is that in many cases the medicine works. It actually makes the person feel much better, so, human nature being what it is, they start to think they no longer need it. And they stop taking the drugs. Then the trouble starts. Add alcohol to the mix and it's an explosive scenario. Had we known about the knife, obviously we wouldn't even have gone in – that's a job for the police – but we just didn't know about it until we walked in there.

A few months later I'm leaving work when I run into one of the coppers who turned up in the convoy.

'I remember you, you got a knife off that enormous Turkish guy,' he says. He then proceeds to tell me he'd written a letter to our boss, saying how he thought Andy and I had done a good job. 'You really calmed him down.'

'Thanks,' I say, making a mental note to tell Andy when I see him. It really isn't the sort of job you want to get too often, is it?

SURVIVOR

Two security staff are waiting for me at the entrance to the shopping centre. They help carry my equipment but remain grim-faced and silent as they lead me along to the back of the building where the men's loos are.

'We had to break the door down to get to him,' says one man curtly. 'So we saw the syringe and needle fall out of his arm. One of the other guys found his bus pass in his pocket, so at least we've got an idea who he is.' He sighs, angry that his otherwise uneventful afternoon has been ruined by a collapsed junkie.

His name is Sean and he's lying in the recovery position on the toilet floor, early twenties, casually dressed, a bit scruffy. Most worryingly, he's a dreadful bluish-purple colour – it's called cyanosis and means he's not getting enough oxygen into his blood cells and tissues. At least he's breathing, though very slowly. But probably not enough to keep him conscious or healthy. Both his pupils are tiny, pinpoint. I contact control and ask for another ambulance. Then I set about trying to help him.

One of the security staff helps me get Sean on to his back and, keeping his airway open by lifting his chin up, I manage to get a plastic device (called an oral airway) into his mouth. This will help keep his breathing passage, the trachea, free from obstruction by his tongue, which could otherwise fall to the back of his throat and stop his breathing altogether.

The guard also helps me by keeping an oxygen mask over Sean's nose and mouth. Swiftly I prepare an injection of Narcan, ready to inject into a muscle in his upper arm to reverse the effect of the heroin. I have to cut his tracksuit sleeve all the way up in order to get to the deltoid muscle where the injection needs to go.

An injection of Narcan into a muscle works gradually – which is what you want to happen. What you don't want in a situation like this is a quick reversal of the effects of the heroin – a rapid return to consciousness can sometimes make the person vomit, in a side-effect of the Narcan. It does the trick. Sean's respiratory rate starts to increase, his pupils dilate and gradually his skin colour comes back to normal. I'm also using the bag, valve and mask device to squeeze in some oxygen to his lungs until he can breathe well enough for himself. It should be quite soon.

The ambulance crew arrive and are starting to get the trolleybed ready to move Sean outside. And yes, the Narcan is working: he is coming round very quickly. He jumps up from the floor and looks around.

'What the fuck's going on, what are you doing here?' he demands. Then he looks at his sleeve. 'What have you done to my bloody clothes? This cost me a fucking fortune, you bastards!' Here we go. Gratitude isn't in it. We've saved his life, he's worried about his tracksuit.

'OK, Sean, you were very poorly, you were barely breathing, we've helped you. Try to calm down,' I say.

'Bastards,' he snaps and tries to walk off, pacing around, mumbling and cursing at us. One of the ambulance crew tries to convince him that it's best to come with them to hospital.

'I've just got out!' says Sean. 'Why would I want to go with you to spend my time in that shithole?' Oh, he's just been discharged from hospital!

'What hospital were you in, Sean?' I query.

'No!' he shouts. 'I've just come out of prison this morning, haven't I?'

Now the penny drops. The chances are, Sean's a long-term heroin user. He probably hadn't been using inside and today, his first day out, he decided to celebrate. He managed to get hold of some heroin and took what he thought was the right amount. But what he didn't realise was his body had been weaned off it. So the dose was too much for him; his tolerance to the drug had dropped and he was even more susceptible to one of its effects, which is to reduce the breathing. Or stop it altogether. Paramedics know that this is an all too common scenario for long-term users who come out of prison.

'Sean, will you let us give you another shot of Narcan to tide you over?' I ask him. The effect of the heroin in his body will last longer than the effect of the Narcan and he could easily collapse again. But, typically, Sean's not having any of it.

'No! Why would I let you do that, you bastards! Leave me alone! You've already fucked up my tracksuit. Get away from me!' he yells. Then he pushes past us and starts running for his life, out of the gents', into the crowded shopping centre. We don't attempt to follow him. It's a far from ideal situation because we know he's still in a dangerous situation. But. as ambulance staff, we're not permitted to treat someone or hold them against their will. We pack up. It's a frustrating experience but sadly one we encounter from time to time.

We can only hope that if he does collapse again, it's where someone can see him. And they ring 999 for help.

DARRYL

Am I pleased that Simon's working with me tonight or what? It's been busy so far, and now we're pulling into a rundown estate, one of the toughest in the area. Simon and I have been good friends for a long time and a call here is never good news: drugs, violence or knives, you name it, this place has seen it all. All we know is a man is unconscious following a disturbance, a fight of some kind. Police are already on the case. Just two people are in the flat: a young man in his twenties and his mother. Neighbours called 999 when they heard them arguing furiously.

'We wanted to talk to him but the guy suddenly passed out in the bedroom,' explains the copper in the hallway. Whatever the situation, Simon and I need to look at him, to check he's OK and breathing, but two other police officers are already in the bedroom, still trying to rouse him. I can hear them through the open bedroom door. They're not having much success. A worried-looking woman, the mum, is in the kitchen, smoking as if her life depended on it.

'I think Darryl's taken some drugs,' she says quietly.

'What kind of drugs?'

'Dunno.'

Now we get the nod to go in. Darryl's face down on the bed, wearing just jeans and trainers. As we put our kit down, amazingly he comes to life, jumps up, flails madly around, sees the assembled uniforms – and heads for the door, hitting me sharply in the chest and pushing me against the frame. The police dive to grab him and he seems to calm down a bit.

'I'm sorry,' he says, then spots me, the lone female. 'Oh no, didn't realise you was a woman, sorry, babe,' he grins. Well, if he's attempting to charm me, at least I can take advantage of it. And he's fine while Simon and I check him out, ask the usual questions. He says he's had a few beers. But he denies taking any drugs. Yet he still seems very twitchy about the police presence and, whenever a copper throws him a question, he's abrupt. And we can see a rather nasty glint in his eye, the kind of glint that says, 'The enemy is here, one move and I'm gonna smash their face in!'

Darry's pulse is racing, a consequence of whatever he's taken. 'I think we should get you to hospital to double-check you out,' I say. But this is a tough estate and Darryl's not having it.

'Nah, don't need it. I'm fine. I've gotta go out and see someone.'

It's past 1am. 'Oh, come on, Darryl,' I wheedle. 'Can't it wait till tomorrow? Your health's more important, isn't it?' He's being quite cheeky with me now, even flirty and I'm playing to his vanity because it could help keep that glint at bay.

Then Simon joins in, 'Come on, mate, it won't do any harm, will it?

He hesitates, looks at us and then at the hated men in blue. 'Oh, all right,' he says, looking me up and down. 'If I'm with you, darlin', I'll be all right.'

Naturally the police offer to accompany us in the ambulance. Judging by Darryl's expression – oops, here comes the glint again – I decide against it.

'No, we'll be fine. You'll behave yourself, Darryl, won't you?'

He grins. I should take the extra precaution but I don't, I'm tired and it's been a long day. For, as I reflect later, his is the grin of a man thinking, 'alone at last'. But I don't see it.

The police then get another call to a fracas very close by and they disappear into their car in a flash. Simon has already gone down to the ambulance, opened the doors and jumped in the back. As Darryl and I go to climb in, he spots Simon in the back, sorting out some equipment. For some reason, he starts behaving as if he's never seen Simon before in his life: another clue I manage to miss.

'What's he doing here?' he says sharply. 'Who's he?' I explain patiently that Simon is a co-worker and a friend, which seems to appease him. But the glint is still worryingly in evidence. To distract him, I start taking his blood pressure and pulse rate again – it remains extremely fast.

'Are you sure you didn't take any drugs tonight, Darryl?' I try again. This kind of persistence usually pays off.

'Yeah, well, I do get depression,' he admits. Then he starts to tell me how he self-medicates with hash and speed, as well as booze. 'The doctor gave me tranquillisers but I dunno, they don't seem to work.'

For part of the journey we chat quite pleasantly. He is flirty, but I assure myself it's harmless. Over halfway there and I'm congratulating myself on having calmed this guy down. Suddenly Darryl undoes his safety belt, swings round and perches himself next to me on the trolleybed. He's right in my face, his arm squeezing my shoulder, pulling me as close as he can. His grip is strong, powerful. Within seconds I'm starting to feel very scared.

'This doesn't bother you, does it, if I hold you like this?' he leers, breathing alcohol fumes at me, his expression now changed from cheeky chappie to menacing groper. Calm yourself, Lysa, I tell myself, my mind racing. Use the right words and it'll be fine.

'Why don't you sit back over there, Darryl, because if Simon stops suddenly you'll go flying.'

'Fuck Simon,' he says with considerable venom. Oops. Stupid of me to even mention Simon's name. Darryl's grip tightens. Now he's staring at me with very obvious sexual intent and I'm briefly at a loss. Simon should be able to see that Darryl has changed seats. But his view from the rear-view mirror is limited.

'Come on then, what about sitting back in your seat for me?' I try in vain.

'I wanna sit here with you!' he says, simultaneously slipping his right hand down to rest on my breast. Instinct, of course, tells me to shove him away, but he's big, volatile and

God knows what substances are in his system. I feel sick to my stomach anticipating his next move. I'm tempted to shout out to Simon but I can't risk antagonising Darryl any further. Who knows what he's capable of?

'I think you're lovely, you're just my type,' he says, his hand squeezing my breast. Now he's attempting to push my head towards his face, trying to kiss me. I'm using all my strength to resist while my mind races, trying to find a way out. We must be pretty close to the hospital, I think. But something has to happen now, Lysa! OK, I'll use the element of surprise. So, just as Simon is reversing into the hospital parking space by the emergency department, I leap up, freeing myself from Darryl's embrace a split second before the ambulance stops.

I fling open the ambulance door. 'They're busy tonight. I'm gonna have to check they have a room for you,' I say, jumping out and running straight into the department, heart pounding. I've been extra lucky. Another few minutes and it could easily have turned very ugly. I'm still shaking when I encounter a security guard and briefly explain what's gone down.

'I'm pretty sure he could be dangerous,' I tell him.

He's ready to radio a colleague to come out to the ambulance when Darryl comes flying through the A&E doors, screaming at the top of his lungs:

'I'm gonna fucking kill someone!'

The complete maniac, a shirtless, snarling beast – totally off his head.

'I'm gonna fucking kill that bitch!' he screams as the security guards run towards him and attempt to overpower him, joined by a couple of paramedics who've spotted Darryl and pile in to assist. There's a big scuffle but they manage to restrain him until the police turn up a few minutes later.

Simon and I watch the police van containing the deranged Darryl drive off. We go back to the ambulance to debrief and take the next call. It's just another job and no bones have been broken – but boy, has it spooked me.

'I should have seen it coming, Simon,' I say as we wait for the next job. 'You could see he was dangerous just by looking at his eyes.' Unfortunately, getting older doesn't always mean getting wiser.

IT SHOULDN'T HAPPEN TO A PARAMEDIC

A PRIVATE MOMENT

'Collapsed behind locked doors.' This kind of call can mean anything. It can be trivial – a concerned friend or relative who can't get any answer on the phone or the doorbell, when often the person is just out of the house. It can be a case of us having to break down a door or climb into someone's home through a window, perhaps to find a fallen elderly person. Sometimes we find a corpse.

Corpses are horrible, especially if they've been lying there, unmissed, for months until the smell alerts someone that something is wrong. In summer the stench of a dead body can be dreadful: heat speeds decomposition and the flies don't waste any time coming in the windows to party on the corpse. Even in winter, if the heating has remained on and windows are closed, it's still pretty gruesome. You can wash away the remains of the smell that seems to linger on you after such a call. But you can't always shower away the images of a decomposing human body.

Today I'm working on my own, and the address is one I'm familiar with, a hostel-type bed-and-breakfast, mainly inhabited by alcoholic men. The caller, Andy, is waiting for me outside. He's looking worried.

'It's my mate, Danny. His door's locked. We usually leave our doors on the latch so we can pop in and out of each other's rooms. I've gone out to the off licence for some beers for tonight and when I've come back, it's locked. You can hear his TV but he's not answering. And he didn't say he was going out.'

The door to Danny's little room isn't the most secure in the world. It looks like it's seen quite a bit of action: the frame has obviously been repaired a few times and the lock is very flimsy. I knock and call his name. Nothing. The television is indeed blaring – but no sign of Danny opening the door.

'Are you sure no one's seen him leave?' I check with Andy.

'No, I checked with the guy on reception. He's in there, I'm sure.'

I'm pretty sure he's in there too, probably unwell, unconscious. Or worse. I contact the control room to ask them to call the police, just in case, and tell control I'm kicking the door in. One, two three, kick! – and it gives straight away. We're in a rather cluttered, untidy and damp bedsit. It's afternoon but the curtains are still drawn. And we can see an outline on the sofa. I open the curtains and I'm looking at Danny, silent, alive and slumped there. It takes the briefest of moments to actually register what's in front of us. It's a bit of a shock, to say the least.

Danny's half-naked. His jeans and dirty underwear are crumpled round his ankles. No wonder he's locked the door. He's been having his own private moment here in this room full of takeaway foils and empty beer cans. But nature – or the parlous state of Danny's health – has intervened. He's conscious but he can't speak or respond in any way. But I can see from his eyes that he's in considerable distress. Then, with incredible comic timing, the long arm of the law turns up and two male police officers barge in. There is the briefest of pauses as they survey Danny's nether regions, his mouldy underpants and the scattered porno show and register what's happened.

'Blimey,' one says, unable to resist the quip. 'What a way to go, eh?' But Danny's not gone. I pull a blanket from the bed over his lap, check him out and see if I can work out what's happened. His body is listing to one side and his face is drooping. One arm and one leg are definitely weak.

'I think he's had a stroke,' I say to the three men in the room. Then, of course, I wish I'd chosen my words more carefully. The men, I can see, are finding this all quite hilarious. I can see the humour too, but I'm sure that despite Danny's condition he can understand what's going on and is deeply embarrassed by the dirty trick his body has played on him at such a delicate moment. So I ask the police officers if they'd mind going outside and letting me get on with helping Danny. Andy wanders out too.

I check Danny's blood pressure, pulse and blood sugar, telling him how we'll get him to hospital, help him, trying to gloss over the somewhat unusual circumstances of his illness. And at least he can now communicate with me through blinking his eyes, once for 'yes', twice for 'no'. So I manage to get a bit of information from him about his medical history, after I've popped a cannula into his arm and taken blood samples for the hospital to use once he's in there. As unobtrusively as I can, I even manage to tidy up the magazines, get them into a discreet pile, as I check the room for any prescriptions or medicines.

Now my colleagues have arrived to move Danny to the ambulance. Outside, I say farewell to Andy, who is promising to come down to the hospital later to check on his mate, 'after I've watched the football', though you know he really means after he's consumed all the cans stashed in his room. But he cared enough about his friend to call us. That's what really counts.

No one actually says any more about the state we've found Danny in. There are a few raised eyebrows and grins but the rest is left unsaid. If Danny makes a good recovery – which I hope he will – he too will be able to have a good laugh about it all with Andy over a beer and a takeaway. You can only hope for the best for him. And, of course, I'm laughing too. Sometimes real life is funnier than anything you can see on television.

LABOUR PAINS

She's outside a pub, sitting on the pavement with a friend next to her. Both are large ladies in their late thirties or thereabouts. We've been called, mid-afternoon, to Norbury High Street: 'Woman, 40 weeks pregnant, in labour.' As we pull up in front of them, they both jump up eagerly.

'Er, that's a bit difficult for someone in labour,' I comment to Gino as we get out of the ambulance.

'Yeah, she did get up a bit too easily,' he remarks.

Then, as we walk up to them, the bigger one, presumably the mum-to-be, looks around and says loudly, 'This will show 'em, won't it?'

'What do you mean?' says Gino. We're already starting to feel suspicious about all this. We expected a woman in considerable distress. This woman doesn't look like she's in pain.

'Well, all those people who think I'm not pregnant. This will prove it, won't it? I can't wait to see my doctor's face when I turn up with this baby.'

Oh dear. Mentally, Gino and I are raising our eyebrows to the sky.

'Who says you're not pregnant?'

'Everyone. And my husband. He says I'm just fat,' she retorts.

He's got a point. She's seriously overweight, with a huge tummy. Could she let us look at her notes, please? Most pregnant women carry their medical notes with them – they're advised to do this in case they suddenly go into labour or need hospital treatment quickly.

'Haven't got any notes, have I?'

'Right. Well, let's get you into the ambulance so we can have a look at you,' I say, motioning to Gino to help.

'They don't believe I'm pregnant,' she repeats yet again. It's like a mantra.

'Well, has your pregnancy been confirmed? Have you had a test?' I venture.

No reply. Without warning, her face contorts with pain. 'Owwww!' she moans, clutching her tummy. 'Ooh! I'm going into labour!' It's almost as if, now she's in an ambulance, she feels honour-bound to create the contractions. It all looks distinctly fishy – and fake. But we can't just walk away. And, no matter how bizarre it all seems, she could still be pregnant.

The friend, a redhead with a tight top and lurid-looking leggings, takes her seat in the ambulance. She gestures to me. 'Even her husband won't believe her, y'know. But I've had five kids. I know a pregnant woman when I see one – and she's definitely having a baby!' Thanks. Funny, the contraction seems to have abated. The friend sits there, grinning at the Pregnant One, complacent in her support.

Gino attempts another tack with the Pregnant One: 'Have you had a scan?'

Quick as a flash, she's got a fresh angle for us: 'Babies can slip behind organs, y'know.' Of course. Every healthcare worker knows that a seven-pound baby can play hide and seek behind your liver. Isn't that what babies do? But I keep plodding away, asking the questions.

'Have you had a pregnancy test?'

Silence.

'Have you had a blood test?'

'Look at me!' she screeches, clutching her big stomach. 'Are you trying to tell me I'm not pregnant? Wait till my doctor sees my baby!'

'He's not the father, is he?' quips Gino, thinking that humour might be a better bet. She laughs. The friend chuckles too.

'Nah, it's my husband's baby,' says the fat one. 'But he says I'm too fat. Do I look fat?'

'No,' I coo. If this is what I think it is, we're better off humouring her.

'I'm not fat, I walk miles every day. This is,' she says, pointing to her stomach, 'all baby!'

We're getting nowhere fast. Then she launches into another 'contraction', moaning and cuddling her tummy as if to protect the unborn baby. Quickly I move to put a hand on to her tummy, a golden opportunity to solve the big mystery. I know that if this is a genuine contraction, it will be rock hard. It's not hard at all. It's squishy. Just an everyday, fat stomach.

'Your tummy's quite soft,' I tell her.

'Pain's gone now,' she says succinctly.

Frustrated, Gino and I just go through the motions, taking her pulse, doing her blood pressure, all the things we usually check. Everything's OK. Now what? How daft will it look if we take her into maternity? The evidence is, this is some sort of phantom pregnancy. An added complication is the fact that she isn't booked in to any unit and therefore has no maternity notes and – most of all – she's not pregnant!

Gino and I leave them inside the ambulance briefly so we can have a hasty discussion. He says our only option is to go straight to the hospital. They'll sort it out. I wonder. Is that all we can do? I've never been to a call like this. And I'm sure no one I know has either. The fact is, it's only my view that she's not pregnant. She needs an internal examination by a doctor to confirm it, for someone else to say, with total confidence, 'You are not pregnant.' And then arrange some counselling for the poor woman. So we decide the best thing is to ring the hospital's maternity ward and tell them our suspicions.

'I've got a woman here who insists she's having contractions but I don't think she's pregnant,' I explain. 'To be honest, I don't know what to do: can we bring her in for you to have a quick look?'

'Yeah, bring her in, don't worry.'

So we drive her in. She sits there grinning all the way. Her friend remains supportive: 'If she's not pregnant, I'll eat my hat,' she says, still firmly convinced by the Pregnant One's fantasy.

We leave them with a midwife and go off to wait for the next job. We're baffled but relieved, another call we won't forget in a hurry. And sure enough, when I ring through to check later on, the news is, the Pregnant One has bottled out.

'She had a really good go at it, she pushed and she pushed,' sighs the midwife.

'But she couldn't get any baby to come out. So she put on her coat and went home in a huff.'

Later, when I think more about this, there are so many questions. I'm sure the hospital would have alerted her GP that there were problems, so maybe she'll get the support she needs. But is this the kind of woman who winds up baby-snatching? Why was her friend so convinced she was telling the truth? And what about the husband? He obviously knew the reality. Too many unanswered questions.

It's funny. But it's terribly sad too. For whatever reason, she really wanted to be pregnant. But she just couldn't face up to the fact that there was no baby inside her.

PASSION IN SURBITON

'Male, 45, leg injury. In garden.' What on earth is someone doing in their garden in Surbiton at 3am? A post-barbie injury, maybe? Too much to drink.

'What else can it be, a burglar?' I say to Gino as we pull up in the silent suburban street. There's a woman standing anxiously outside. She's attractive, deeply tanned and skimpily clad in shorts and t-shirt on a warm summer's night. Before we can say anything she grabs Gino's arm. 'Can you come down the side entrance and be really quiet, please?' she whispers. 'We can't wake anyone up.'

'Is this your house?' whispers Gino.

'No. I live there,' she says, pointing to the house next door.

Hmm. Maybe she's spotted the injured man from her window and is doing the good neighbour bit.

As quietly as we can we go through the side gate into a 100-foot-long garden. At the bottom there's a sort of play area with a slide and a swing. A dark-haired man in his forties is sitting on the grass underneath the swing frame. He's wearing just tracksuit bottoms. And straight away we can see he's got a nasty fracture or dislocation. His foot is pointing at an odd angle. He can't move. And he's struggling with the pain.

'You've gotta help me,' he says in a loud whisper. 'My ankle's killing me. Can you give me something now for the pain?' Normally we'd have a quick chat to find out how he'd managed to get this way. But neither he nor the woman seems willing to give us much information.

'What were you doing?' I whisper. 'Were you trying out the swing?' I say, hoping to get a laugh from one of them. But she says nothing. There's an awkward silence.

'Ah… I dunno. It just went like this all of a sudden, it twisted,' he says, wincing at Gino. 'Have you got the painkillers then?'

This is all very odd. Normally in a situation like this you can't stop people from talking – they're usually only too keen to give you their story. These two, however, are suspiciously quiet. We give him some gas and air for the pain and Gino heads back up the path to get the carrychair and a splint – there's no way this guy can walk.

I carry on whispering the usual medical questions and get brief responses from the man until Gino comes back and we get the splint on to his ankle to stop it wobbling around. We're working as quietly as we can. Then we get him into the chair and up the path into the vehicle.

Once inside the ambulance, there's the usual paperwork. The woman says she wants to come with us. Fine. 'Is this your address?' I ask the man.

'Er, yes.'

'What number is that?' I ask, gradually twigging to what's been going on here.

He gives it to me. OK. 'Is this lady your wife?'

'No.'

'Can we have your wife's name and number, please?'

'You're not gonna ring her, are you?' he snaps.

'No, it's just for our records as next of kin.' Grudgingly he gives it to me. Oh dear. As we drive off, the woman is now staring very pointedly out of the window. She's mortified. They don't exchange a word to each other.

Hanky-panky in Surbiton, eh? In the dead of night, these two have both been climbing out of their respective beds, leaving their snoring spouses, while they indulge in a nocturnal lovefest. You have to hand it to them: it's pretty close to home but it is a secluded spot. Clearly their partners don't suffer from insomnia. As for the swing, well, take your pick as to what they've been trying to do! Gino and I have a good laugh about it after we leave them in the hospital. He'll have to tell his wife something as he will probably need an operation quite soon.

'How's he gonna explain an injured ankle in the middle of the night?' chortles Gino. 'If he got into bed with the wife as usual, it's not looking good to suddenly turn up in hospital, is it?' We continue to speculate, until Gino goes off for coffee. I'm still in the ambulance, writing up my notes, when I see the woman outside the hospital, nervously puffing on a cigarette. She spots me, stubs it out and walks over. I wind down the window.

'Can I just talk to you for a minute?' she says. It looks like she's been crying. I jump out to talk to her.

Obviously she wants to unburden herself to someone and she tells me the whole story. They've been having an affair for over two years. She's madly, desperately in love with him. Neither couple have any children. She's been very unhappy in her marriage for years. But he's adamant that, while he loves her, he doesn't want to divorce his wife. Sometimes, she says, they'll manage to sneak off to a hotel when their partners are away on business. But in the summer they'd risk it out in the garden.

'I'm sorry we got caught out like this but I'm almost hoping that if his wife does find out now, it'll bring it to a head,' she tells me hopefully. 'I know he's the love of my life.' I make sympathetic noises, but what can you say in a situation like that? Give the guy up? Don't risk it in the garden? Stay away from swings? Go to marriage guidance counselling? I've no idea what went down afterwards. But Passion in Surbiton will live on in my memory for a long time to come.

A SCRATCHY END

A really hot night, early summer. One of those rare balmy English summer nights when everyone's in a good mood. In winter most of our jobs are cars going into each other or pensioners falling over at home and breaking hips. But in summer we usually expect to be called outdoors to accidents with barbies, DIYers falling off ladders or weekend sporting accidents. So when we get a call to a 'Body seen on railway embankment' Liam and I look at each other and start speculating.

'Someone's jumped off the bridge,' says Liam. Like me, he has been in this job for years, and he's good value as a colleague.

'Yeah, but it could be someone gone in front of a train,' I offer. 'Could be that the impact's thrown them.' It could turn out to be a really gruesome job. But the reality is, we don't really know what to expect. A person could be dead or half-alive and mutilated – anyone's guess.

We're directed to a railway bridge near Crystal Palace. But it's not as easy as we thought to actually get on to the railway embankment. After swinging off the main road we waste precious minutes trying to find a road that takes us closer. No joy. So we park the vehicle and stop to ask some kids passing by on bikes. They walk us along the road and show us the fence that runs along the side of the track.

A quick leg-up from Liam and I'm going right over the fence without a problem. I love climbing. I'm like the proverbial rat up a drainpipe. But this is quite a high fence and as I start to lower myself down I realise that underfoot it's all prickly and uneven and it slopes down towards the track. Once I'm on the ground, I can just about see the bridge. But I'm already waist deep in brambles and nettles – and they bloody well scratch. Comically, I'm having to lift my knees up really high just to get through. The police are massed on the bridge. I shout over to Liam that he should go and join them.

The overgrown greenery is even deeper than I thought. It's not just scratchy, it's thick. And, wouldn't you know, I can't see anything like a body. But wherever it is, the coppers on the bridge seem to think they can see it. Now they're yelling at me, telling me which way to head – using vigorous sign language too. Liam has managed to tip my heavy kit bag over the fence, which doesn't make it any easier. And now I'm in pain and it's not funny any more. I'm suffering death by nettles. They're everywhere and I'm getting scraped and torn by the brambles going right through my trousers. It really stings. This is turning into a bit of a fiasco. And, of course, I've got no idea what I'll find at my destination.

'Over to the left!' I hear from the bridge, the guys peering at me as I struggle to hack my way through. I'm sweaty and pissed off but I can see how funny it must look: Lysa struggling to stay upright while being scratched to death. I grit my teeth: the job's got to be done. And at least it's still light. I hack through a really scratchy patch, only to fall over – pure stand-up comedy for the onlookers, who are having a good laugh at the spectacle.

Oh dear. What will I need to do if the person is still alive? Will we have to get the fire brigade to help us? I wonder if it'll be me who has to rouse them from their life of leisure? It's a running joke between the ambulance service and the fire brigade that we're much busier (and it's true, you do a great job, guys, but you don't handle as many emergency calls as us).

So there I am, knee-deep in nettles, providing the evening's free entertainment. As I clomp round a tree, I'm struggling to move uphill again. Then I spot some plainclothes coppers on the bridge. That means CID.

This is no joke, I tell myself. What the hell am I going to find? Either a body or a psych patient who tried to end it all by jumping off a train but made a hash of it. I reckon I've been hacking my way through the jungle of this embankment for about twenty minutes. All I can see is rubbish everywhere: battered cans, used condoms, fag packets and the rest of the detritus of many years of not-very-environmentally-aware travellers passing through. At least there are no trains. But if it's a naked, dead body – which is what the original call-out guys seemed to think it was – how could they lie down in this by choice? I'm not nervous about finding a corpse. That's part of the job for us. But I'm more concerned about what sort of condition the dead body might be in. That's if it's even intact.

This is ridiculous. You sign up to save lives and you wind up looking for dead bodies on railway embankments. I start working my way uphill towards the body. And oh, now finally it's there in front of me. I can see it. It's lifeless, all right. In fact it's never known the joy of being alive and being scratched to hell on a hot south London night. All our efforts, police, 999, CID, have been directed towards the rescue of a naked shop dummy. Someone probably chucked it over the bridge – maybe even the work of those kids on bikes.

'It's a bloody mannequin!' I yell through clenched teeth.

And, predictably, everyone on the bridge falls about laughing. It's a big relief really. And I'd probably be laughing too – if I didn't have to make my return journey through the brambles. There's an even higher bit of fence to scale on the way back. But I can manage – sometimes it does come in handy having the words 'Circus trapeze artist' on my CV. So I find some concrete uprights holding the fence and scramble up and over.

Liam, of course, isn't in the least bit sympathetic as I moan about my scratches and the nettle cuts on my legs. 'Nothing was gonna stop you getting over that wall, girl,' he reminds me. But still, I reflect, as I write it up afterwards, no harm done. A bit of a laugh really. And let's face it, we can all do with that sometimes.

NO SEX IN THAILAND

Do a job like mine and you'll probably wind up with a nickname. Mine, as I've referred to earlier, is Barbie medic. And no, I'm not offended, a joke's a joke, after all, and being five foot eight with blonde hair does have certain advantages. But there are times when I manage to live up to my nickname. Like tonight.

I'm making myself look a right fool because I can't quite make sense of what control are telling us. 'A man has shot himself with a slug pellet gun.' Now, I'm a keen gardener but I'm a bit stuck here. Why on earth would you need a gun to kill slugs? They don't move very fast. You could just throw the pellets on the ground. Brian in the control room doesn't know what I'm on about.

'Why do people need these gadgets, Brian? It's so silly,' I say. 'I just chuck mine on the ground.' Brian doesn't answer. He's too busy. But my co-worker tonight, Ian, sets me straight.

'You stupid cow, it's not a gun for slugs in the garden. It's a gun that shoots pellets called slugs, geddit?'

Oops. So this is potentially quite serious. The road's already been cordoned off because there's a gun involved. It's classified as an armed incident. So we whip out the stab vests, though we're still being told to wait at the end of the road. The police can't let us go anywhere near the place until the person has been disarmed and it's safe. Having said that, we're fairly sanguine about what we'll be up against. Many armed incidents come to very little: either it's a hoax or there's no trace of a gun.

But this time it's not a hoax. And the police are now giving us the OK to come through. There's the man who's shot himself, sitting on a wall outside the block of flats, puffing on a cigarette, a can of beer beside him. He's a pretty nondescript man of about fifty with an Eastern European-sounding surname.

'Where are you injured, then?' I ask him, slightly relieved to see no obviously nasty signs of trauma. He gestures to his mouth. Sure enough, with the help of my torch, I can see the blood on his upper palate.

'I shoot myself,' he informs me. He seems quite proud of it.

'Why would you go and do that?' I ask.

'The wife,' he replies. 'A waste.'

'Waste of what?' I ask innocently.

'I want sex. She doesn't give me sex. She my wife, she should give sex every day. When I want it. I tell her, "I want sex, you don't give me, I shoot myself." So I shoot myself. Two time.'

No nookie on tap, so hubby shoots himself. Good job half the men in south London don't resolve this little domestic problem in the same way or I'd be on overtime for the next ten years. But as I examine him more carefully, I can feel a little pellet. It's lodged under the skin at the back of his neck.

'We have to put a collar on you and get you to hospital,' I tell him.

'No. I all right,' he insists.

'Having two bullets lodged in your head is not all right,' I counter. But I've only found one. I can't for the life of me figure out where the other one has gone.

'Things could get very nasty. You could get an infection. Or the bullet could be lodged too close to your spine.' He's not troubled by the risks he might be facing. He's already had a good look at me. And he likes what he sees.

'Are you married?' he leers. He's no George Clooney. In fact he's probably high on the list of the least desirable men in south London. But there's no question he fancies himself as Jack the Lad with the ladies.

'Yes, I'm very married,' I tell him, thinking that might shut him up. Of course, that goes right over his head. Now he's giving me the full blast of his charm.

'Ah, a nice English wife. Better than my wife. She from Thailand. No good woman, Thailand. Not interested sex. But English girls are,' he adds. Well, he can live in hope.

By now the police have had a chance to talk to the 'no good' wife in the flat. And they've retrieved the weapon, thankfully a low-powered air rifle.

'You give me your phone number,' orders Mr No Sex.

'You've got it, it's 999,' I say, our bog-standard response to what is, funnily enough, a common scenario. It goes with the job. All the other female paramedics and nurses I know agree: a man can be the dirtiest, smelliest, most uncouth individual on the planet, but he will convince himself he's got half a chance when you're looking after him, even at a time when he might be on his way to heaven.

Time after time we hear: 'Any chance of your number, darling?' It's pretty harmless. It's as if they have to assert their masculinity in their time of trouble, to distract themselves, perhaps, from their frailty. Surely Mr No Sex can see he's definitely not on to a winner? But no, he continues his charm offensive. No wonder his wife got fed up.

'You good-looking woman. You like sex?'

I ignore this. He's going too far.

At least all this nonsense changes the mood. It's amazing how the prospect of an encounter with a strange woman – no matter how remote the guy's chances – can lighten things. Now he's established his credentials in the overwhelming sexual allure department, he's changed his mind about going to the hospital. He obediently lets us put a collar on him and get him into the ambulance.

Strapped to the trolleybed, Mr No Sex pursues his goal. Would I go for a drink with him one night? Do I like dancing? Where do I live? He's harmless enough. I can't be sure but I think there's a bit of a problem here, maybe mental illness. But tonight we don't get to follow up after we get him to hospital. And I don't even get to see the wife. She doesn't come with us in the ambulance. Which probably speaks volumes about the state of their relationship. Not much love lost between them.

'Those two weren't getting on too well, were they?' quips Ian as we get ready for the next job. Says it all, really.

CORRIE

In the old days we would rush to a call and get the patient to hospital every time. You'd do your best on the spot – but essentially we got there with the aim of handing over to hospital staff.

Now, after some 999 ambulance calls, a person doesn't need emergency hospital treatment at all. We can treat them there and then. I may know in advance that a call is low-priority and I can wind up helping someone in their home, maybe even having a cuppa and a chat with them before leaving for the next call.

Now and again you get a really memorable low-priority call.

An elderly lady, Ruby, has accidentally cut the front of her leg. It's the kind of injury that can be quite common when you get to your seventies and eighties. The skin becomes very fragile and thin later in life, so it may tear or be damaged at the slightest knock.

It's been a quiet night as I drive into one of the area's poshest roads to see Ruby. And it's a wow-factor house, big, detached with a stunning garden and its own driveway. Ruby's daughter, Claire, is all smiles when she takes me through to Ruby in the big lounge. Teenage boys are slouched in front of the TV. On seeing me, they switch it off and disappear upstairs. Ruby's alert, silver-haired and in her early eighties. Her eyes widen as I walk in.

'What have you been up to then?' I ask.

'It's you!' says Ruby, grinning broadly at me as I set my bag down.

'Yes, it's me,' I chirp, thinking she's probably a bit confused as Claire explains what happened. Ruby was quite good on her feet but tonight she managed to catch the front of her leg on a coffee table. She had a nasty laceration which had obviously been bleeding quite a lot.

'Will you take her to hospital?' says Claire nervously.

'No, I can sort it out here,' I say confidently, opening up my kit. On a wound like this, you don't use stitches because the skin is far too fragile. It's a case of cleaning it up, gently pulling the skin flap back to near normal and taping it, then putting a dressing and bandage on top.

'All you'll need to do when I've finished, Ruby, is keep your leg up when you're sitting down,' I explain, getting to work on the wound and hoping fervently that Claire will read my mind and suggest a cup of tea. It's 8pm and I haven't had anything since lunchtime.

'Look who it is, Claire!' says Ruby, now visibly excited and grinning like a maniac.

'Yeah, Mum, it's the paramedic, I know,' says her daughter.

'No, you dummy, can't you see who it is: it's *her*, from *Corrie*, from the Rovers, it's Raquel!' I look up at Claire with an 'Is your mum all there?' sort of half-smile. I've no idea how confused Ruby is or what her usual frame of mind is. Older people do sometimes mix you up with someone else. I've been taken for someone's relative before now. I've even been mistaken for a man (it's the uniform, I hope!). But a TV barmaid? This is a novelty.

'I'll just put the kettle on,' says Claire, motioning me outside the room to give me the background to her mother's case. Ruby has been living with Claire and her family for over a year now. She's pretty healthy and quite mobile. But she does get confused sometimes.

'One night she went out in the street in her nightie and the police found her and brought her back,' says Claire. 'It's probably better if you go along with the Raquel thing. She loves *Coronation Street*, never misses an episode. And she won't have it if you tell her it's all in her mind. She'll get really upset. To be honest, by the look of her, it's really perking her up. I haven't seen her so happy in ages.'

Fine. We get our fair share of abuse from the public, so it's nice to step up in the world as a soap queen, albeit one who left the show years ago. We're trained to try to orientate confused people into the here and now but a bit of harmless impersonation never hurt anyone, did it? I go back in to finish bandaging Ruby's leg.

'I'd better be on my best behaviour, with you around,' chuckles Ruby. 'Wait till I tell them at the daycare centre!'

'Claire!' she hollers out to the kitchen. 'Bring all my pictures from upstairs!'

A few minutes later Claire comes in with tea, biscuits and Ruby's *Corrie* memorabilia. There are pictures and magazines all in a big box. It's a repository of decades of dedicated viewing of the nation's favourite soap. Pulling out a faded listings magazine from the nineties with Sarah Lancashire as Raquel on the cover, Ruby beams at me.

'Give us your autograph, Raquel, so I can show everyone at the centre.'

I oblige, of course. As I try my best to scribble something – I can't quite remember if it's Sarah Lancashire or Sarah Lancaster – Ruby lights up with happiness. She's obviously forgotten all about her leg. 'After all these years,' she says contentedly, 'I've met Raquel. Now I can die happy.'

'I don't think you're going anywhere, Ruby,' I say, finishing my tea and packing up. 'I think you'll give us all a run for our money.'

'Look,' says Ruby, proffering the plate. 'Take these custard creams with you when you go back to the pub. You need to keep your strength up. It's a disgrace, the way they treat you.'

'What's that then, Ruby?' I smile. She's looking really concerned now.

'Well, you have to work all these hours, in and out of people's houses. And then you have to go and work behind that bar to make ends meet! They don't pay you enough, do they?'

'No, Ruby, they don't,' I agree. 'But thanks for the custard creams.'

AND ANOTHER THING...

REACTIONS

One of the tougher aspects of the job is dealing with the emotions that relatives or families go through when a loved one dies unexpectedly. It might sound surprising but most people are quite stoic and restrained. Despite what you see all the time on TV or in the movies, generally speaking people don't scream or go into uncontrollable hysterics. Of course there are tears. But not every time. Perhaps these people save their emotional outpourings for a private time or place, I don't know. But even though witnessing other people's grief and shock does come with the territory of emergency work, there are times when you can only shake your head in wonder at people's extraordinary reactions.

Confusion. Bob and I are on our way to help another crew. We're told they've gone out to a sixty-year-old woman with a leg injury. When we're almost there, the same address comes through again. Only this time it's a thirty-year-old woman, cardiac arrest.

'Bit of a communication breakdown there, Lysa, eh?' says

Bob. The door to the terraced house is wide open. In the front bedroom two paramedics, Steve and Mike, are busy carrying out CPR on a young woman lying on the floor by the bed.

There's also an older woman, Doris, sitting up on the bed in her nightie, seemingly oblivious to what is essentially a knife-edge, life-or-death drama being carried out in front of her. It's so odd. In a situation like this you'd normally expect some display of emotion, concern or even fear of the worst from a watching close relative. Not Doris. Her knee is bloody yet she's smiling and chatty.

'Hello there,' she chirps to me and Bob as we walk in. 'It's getting really busy now, isn't it?' We acknowledge her briefly but the priority is, of course, the person on the floor. Steve, one of the two paramedics involved in trying to resuscitate the younger woman, very briefly fills us in. They'd encountered quite a different situation to the one they'd been expecting.

'We got here expecting to help Doris here with a leg injury,' says Steve.

'Instead, we found her daughter like this.'

Claire had been bringing Doris her morning cup of tea when, without warning, she suddenly went into cardiac arrest and collapsed. She fell heavily. And in falling, with her mouth open, she'd somehow taken a chunk out of Doris's shin with her teeth.

Ambulance staff are an amazing breed of worker in that nothing – and I mean nothing – ever fazes them. They might get sent to a routine call, someone with 'tummy pains', and wind up delivering twins. All in a day's work. So there they are now, on their knees, doing their level best to resuscitate Claire – but tragically getting nowhere.

It's a hot day and resuscitation is hard, physical work. They've been trying for some time, waiting for us to arrive. By now we all know it's a worst-case scenario. Bob and I take over nonetheless. We do everything we can – but Claire is gone. We decide that Steve and Mike will carry on doing CPR in the ambulance all the way to hospital – but nothing short of a miracle will bring Claire back. How awful for her mother to have to watch all this, I think. And how will she cope when she learns the truth?

I'm worrying for nothing. As Steve then starts to help Doris and clear away the blood from her leg wound, she's blithely chatting away to him, as friendly as they come, asking him the most innocuous questions, the sort of chit-chat you might have with a stranger on a bus: 'Do you live near here? How many children have you got? It's still raining, do you think it'll ever stop?'

It's hard to believe that just inches away, two ambulance crew have been desperately engaged in a fruitless battle for her daughter's survival – a battle that has just been lost. There's no sign whatsoever of grief or shock at what's happened. She hasn't even been asking them any questions. She just seems happy to have some company, someone to talk to.

I can't believe this. I manage to mouth to Steve, 'Does she realise she's dead?'

Steve nods. At some point, before we've turned up, they'd

attempted to explain to Doris that technically her daughter was dead. They would do everything they could. But they thought it best to warn her. Yet it seems the truth has not sunk in. Bob and I take Doris to the hospital. I try my best to be subtle as I attempt to gauge her understanding of what's happened to her daughter. She's still relentlessly chirpy and smiling. For whatever reason – though I'm now starting to suspect she has serious learning difficulties – the reality of the situation continues to elude her. Claire, for her, is still someone in the present tense.

'Yes, dear, Claire is such a good girl, she does everything for me, the shopping and most of the cooking,' she tells me and I think how sad it all is. Clearly Claire has been her mum's main carer, her major support in life. How will this woman cope now? I will never know because we take her into the hospital and are quickly called out on to the next call, another emergency. For all I know, she never will fully acknowledge her daughter's death.

Strangely enough, I find myself encountering something similar -- and even odder – about nine months later. I'm part of a crew that have turned up at a house after a body of a young man has been discovered. He too had a sudden cardiac arrest. But it was a very unusual, undetected heart defect called long QT, a heart abnormality that causes a disturbance in the heart's rhythm. It's a swift and painless end – he wouldn't have known a thing. There's nothing that can be done for him. But as the news spreads by mobile phone, the house begins to fill up with relatives, many of whom live nearby. Each relative arrives, looking appropriately downcast and concerned, and then joins the other relatives in the house to talk and console each other.

But then something very strange happens. Somehow, after they've talked it all through with each other, they have developed a strong belief that the man isn't dead at all. They seem quite cheerful, hopeful, almost as if they're anticipating a positive outcome, even though the man's corpse is lying in another room. It's the weirdest thing. Try as we might, they are also refusing to come and look at the body. Instead they assemble in the lounge and start to pray for his resurrection. I go into the room to talk to them because I need some information from them, a GP contact at least.

'Can I talk to you for a moment, please?' I ask one relative, a pleasant-looking woman who might have been an aunt.

'Look, we're going to pray,' she says. 'He's not dead. Just let us know when he comes round.' I'm amazed. This is collective denial. Their beliefs are their beliefs and you might expect some people would to want to pray at a time like this. But they really aren't getting it. I leave them for a while, then I go back into the room again. The man is dead and the formalities have to be dealt with at some point. This time all eyes in the room turn to me.

'Is he alive yet?' says the pleasant-looking woman.

'No, he's still dead,' I hear myself say without a trace of sarcasm. It's at least another hour or so before one man is finally persuaded to leave the room and confront the reality. I've never experienced anything quite like this before – and my colleagues say the same when I tell them. But there are times in this job when you simply cannot fathom people in any way. No matter how hard you try.

REGULARS

Like most other organisations, we have our regular clients. Some of them are so regular I'd be surprised if the phone company hadn't offered to put 999 on their friends and family discount list. We're on first-name terms with many of our regulars (or frequent flyers as they're also known) and it's not unusual for us to know their family and their life story. Some of them know us so well we expect an invite to a wedding, bar mitzvah or christening any day. Why do some people keep calling us out? There are many reasons.

Some, of course, have chronic health problems and frequently need us either to treat them on the spot or get them to hospital if things get bad. Sometimes these people can arrange, with the cooperation of their doctors, a list of specific treatments they might require. This kind of list can save a lot of time if we're called out to them, especially if they've got an unusual condition or illness. If they need treatment at a specialist hospital, this too can also be included in the list.

And then, of course, there are people who are lonely or isolated in the community. Such people may have a lower threshold for dialling 999 than your average person in the street – even if there's not a lot medically wrong with them. And many of these are older people living alone. It's not unusual to be called out to someone like this and realise that you are the only human contact the person may have had for days at a time. And even once we've established that there is no medical reason for us to be there, they will often try to entice us to stay, plying us with offers of tea and biscuits. They may not be ill, cold or hungry. They're just starved of human contact.

Typical is Marge, a lovely lady in her eighties. Her husband died a few years back. She has survived while one by one her friends have also died. Her two children live far away and visit once or twice a year; settling their uneasy consciences by paying for carers to come in to tidy up and shop for Marge twice a week.

Marge is registered blind, so TV or reading aren't really options. She also has tinnitus, which means listening to the radio isn't easy. Unable to get out much, she sits alone in her armchair day after day, week after week. Imagine that. So, not surprisingly, she calls us now and again. She knows that we will always turn up and have a little chat for a while. Who can possibly mind that?

Other regulars can prove more challenging. Like the people who drink the recommended weekly alcohol allowance in one day – or less – but who also have mental health problems. That complicates the picture quite a bit. They're unhealthy or unwell because of their drinking and their behaviour would try the patience of a saint. As their friendships fall by the wayside and they become increasingly lonely and desperate, they reach out to us. Sometimes 999 may be the only number they can dial out because the phone has been cut off for non-payment of bills. So we turn up to find someone behaving in a difficult, even obnoxious way. And we then have the cyclical conversations:

'Come to hospital.'

'No.'

'Oh, go on.'

'No.'

'OK, we'll be off then.'

'I'll go, then.'

'Come on, then.'

'I've changed my mind.'

A few callers in my area have been known to call 999 six times a day – sometimes more. We know perfectly well that when we get there we'll be going through the familiar routine of trying to convince them to come to hospital with us. We know it's going to be a futile exercise. And we know it's often likely to be accompanied by swearing and abusive threats. We also know that A&E isn't really the right place for them and that, in all probability, if we do manage to get them there, they will walk out hot on the trail of the next drink to stave off the DTs. By the time we've finished our paperwork they're glugging the next drink.

This is a cycle that is all too familiar to ambulance colleagues and staff of A&E departments up and down the country. It's a game we play daily, always concerned that the time we leave them will be the time they keel over and die. We might be the last person to see them alive and questions will arise, like, did we do enough?

Psychiatric services are similarly frustrated by this type of caller because they fall between two stools. They may very well have mental health problems, but psychiatric staff aren't there to care for people who abuse alcohol. They'd prefer to treat them while they're sober. And, of course, the drinkers don't get sober because it doesn't feel good to them. So they walk out of hospital with the beer-detector radar switched on to full power and the whole sorry cycle starts all over again.

Sometimes, though, a regular will make a breakthrough. There's one guy in our area we've been going out to regularly for many years. Brian can be sweet enough if you catch him on a good day and I've always managed to achieve some sort of rapport with him. However, he's got a bit of a soft spot for me. He's been known to tell many of my paramedic colleagues (including Steve, my husband) that we're in love, that I 'pop round a lot' and that we share many a private moment together over a can of Stella or twenty. Steve finds it all quite amusing and plays along if he's called out to Brian, never mentioning that I'm his wife. (Couldn't he be a little bit jealous? I ask you!)

Last Christmas I was at a staff party when someone told me my 'virtual boyfriend' had died. I felt unexpectedly sad. No one seemed to know the source of the story, though, or have any details to confirm it, which seemed a bit strange to me. So, next time I had a quiet moment between calls in his area, I popped round to Brian's place and knocked on the door. He came out and looked quite surprised to see me.

'I didn't call, honestly, Lysa, it wasn't me this time,' he said.

I laughed and told him about the rumour. 'No one's seen you for ages, Brian.'

'No,' he said. 'I haven't had a drink for months. So I've tried not to call you lot out and waste your time any more.'

He then invited me in to meet his girlfriend. There they were, watching TV, eating biscuits and drinking cups of tea – not a beer in sight: truly a scene to behold. Naturally, back at the station, we were all delighted. Cynically we'd not considered the fact that Brian might have stopped calling because he was off the booze. It was really good to see him doing so well. Unfortunately the relationship ended – and so did his resolve to stop drinking. And so Brian was reinstated as our five-star, award-winning, regular caller.

THINK ABOUT IT

A while back, some of the more ridiculous time-wasting 999 callers were broadcast to the nation on radio and TV. The woman who rang us to ask what day it was or the distraught woman in her car making an emergency call because she couldn't find Homebase were unfortunately pretty typical of some of the unbelievably daft calls we get. People think we're an all-purpose free public service. You remember the number so you can ring it.

I don't want to be responsible for putting anyone off from calling an ambulance in a genuine emergency. Older people and those with chronic long-term illnesses often don't want to burden the service because they have heard on the news how stretched we are. They don't want to be considered to be time-wasters. When they finally do call us for help, sometimes they've let things get really desperate. I've been to patients breathing their last and their partner says, 'I'm really sorry to have bothered you, I know how busy you are.' When that happens, your heart sinks.

It has to be said, however, that we've all been called out to some very daft calls. I would never put myself in the position of judging who needs the services of an ambulance. And we try our best to treat everyone with patience, dignity and respect, whatever their condition. But afterwards, when I reflect on an unnecessary case, I find myself thinking, Why, oh, why did they call us for that?

Parents of young poorly children can find it very difficult to distinguish the symptoms of a nasty cough and cold from something more severe – especially when they think it could be meningitis, which is quite understandable. I know I might panic and lose the ability to think straight when my own kids are unwell; it's human. But there are often times when you wonder where people's ability to think rationally and help themselves has gone.

Consider the man, intelligent enough to hold a full-time job, who insists he must travel to hospital by ambulance for a check-up. He has nothing more than the symptoms of a common cold. He's also convinced that having been taken in by ambulance, he will receive high priority – a commonly held myth, believe it or not. What he eventually gets is a ticking-off from the doctor. Hospital A&E teams can get quite shirty about this type of thing, especially when it's busy.

The young student with a tickly cough. As sick as she is, she manages to get the bus from her flat to her friend's flat, which is just around the corner from the hospital. Then they make a cup of tea. And then they decide to call an ambulance. I ask you!

Then there's the young woman whose partner rings us. The call comes through as 'Laceration, severe haemorrhage', so we tear through the streets, blue lights and sirens on, to get there. The boyfriend is standing in the road, doing what we call a windmill – arms waving wildly around. Everything points to a genuine emergency as we race, with our equipment, to help her. And there she is, in her bathroom, with a small piece of toilet roll round the tip of her finger. This, then, is our haemorrhaging woman. It's a paper cut. Two fully trained paramedics, in a fully equipped frontline ambulance, who might have been needed for a real life-threatening incident, have been deployed for this.

'This is a tiny injury. You need to think before dialling 999,' I warn her.

'Don't you carry plasters?' she snaps.

'It doesn't need a plaster. And if you do, there's a chemist a few doors down.'

Now she's contrite. She panicked, she says. 'It was when I saw all the blood.'

Consider, too, the healthy woman who has gone into the early stages of labour. She walks into the ambulance unaided and spends the journey laughing and chatting about baby names for the mile or so to the local hospital. Then she walks off. She reminds me of the pregnant woman who calls us out even though she lives opposite the car park of the maternity unit. She strolls with me across the road to the unit while my colleague turns the vehicle round to get on the right side of the road because it's quicker. But at least they're pregnant.

One of my most memorable time-wasters is a forty-four-year-old man complaining of dizzy spells. Then he reveals he's already been to see the GP. Oh dear. Why is he calling an ambulance?

'So what did the GP say?'

'Oh, he gave me these tablets,' says the man, producing a pack of Stematil, an anti-sickness treatment for dizziness.

'So when did you take them?'

'Oh, no, I haven't taken any.'

'Why not?'

'Well, on the box it says, "Take every eight hours." And I haven't had the pills for eight hours yet.'

Last but not least on my hit list of top time-wasters are the alleged good Samaritans who drive at speed past some poor old boy having a nap on a bench or something similar. Not concerned enough to pull up and double-check, they whip out the mobile to report someone in cardiac arrest, though they've usually only the vaguest idea of where they actually saw the person! I've lost count of the number of times me and my colleagues have woken someone from a pleasant afternoon siesta in a sunny spot as a result of worthy, caring citizens deciding to call 999 because they're too busy to take a few minutes to find out the truth for themselves.

TIPS ON CALLING 999

Ambulance teams are there to help the public – but there are a few things the public can do to help us. You might find you only need to call us once or twice in a lifetime. Or indeed you might never call us at all. But if you do have to call us, there are ways of making our job just a little bit easier.

It's a big help if you give us as much information as possible about your address. It's amazing how many people there are in London who don't have clearly numbered addresses – and this applies to business addresses too. So if you know your address is hard to find or tricky to access, give us a few clues. Maybe there's a road junction close to your home, or a nearby landmark. Even giving us the make and colour of a car parked outside your premises can help us. And, of course, if there is someone with you when you call us, it helps if they can come outside and wave us into where we're needed.

I was once asked to go to help someone in a flat for what sounded like it wasn't a life-threatening incident – which was just as well. I found the road and drove up and down. But I couldn't find the specified block of flats. I called our control room and asked if they could get back in touch with the caller to get more detail. It was next to another flat apparently: very useful. What was more, the caller couldn't come out to meet me. I parked, grabbed my heavy bags and set off to explore. I asked several passers-by if they could direct me to the flats. Eventually someone was able to set me straight: yes, there was a block of flats backing on to the road, but the actual entrance was in a different street. When I eventually arrived I suggested to the patient and his wife that it might have been really useful had they provided this information when they made the call.

'Yes, it's funny you should say that,' said the wife. 'We always have problems when people try to find us. Hmm. Maybe we should think about doing that in future.'

Another thing that is important for us, once we've located you, is access into the home. We often ring the doorbell to hear, 'Who is it?' They've just rung 999. Who do they think it is?

We say it's the ambulance service. And then we wait on the doorstep. And we continue to wait. Until bolts are drawn and a series of keys eventually unlock the door. A few minutes later and we're inside the residential equivalent of Fort Knox. So to say that we appreciate it if the front door is left open – or at least unlocked – is an understatement. It saves time. And for the emergency services, time is the most precious commodity.

Another thing that helps us a lot, if you're waiting for an ambulance to arrive, is if you can move your pets into a different part of the house. No matter how adorable they are, sometimes they do object to our presence, particularly if we have to treat their much-loved owner. My husband, Steve, was once bitten by a dog protecting its owner. It was a lovely dog. But he was scared and didn't understand what was going on.

If there's time, do a brief note detailing your GP, regular medicines, any allergies and name and details of next of kin. It's not essential but it can be useful to us.

Don't move an ill friend or relative too much unless it's obvious that they are in danger. It's also not a good idea to move someone to the highest point in the house. If they're already there, that's fine. But it doesn't make sense to move an elderly person – or, come to that, a heavy, muscle-bound rugby player with a leg injury – up three or four flights of stairs, because we are going to have to carry that person down the stairs in the carrychair. Think of our poor aching backs!

Our job isn't just about getting into the address. We spend a great deal of time on the road, getting through the traffic. So other motorists too can make our life a bit easier. Here are a few guidelines for drivers:

If you're driving along and hear the sound of sirens, have a quick check around you to establish where those sirens are coming from. It's usually best to indicate and pull over to the left. We tend to like to take the middle of the road, rather than weaving our way through the traffic. This reduces the potential hazard of hitting a pedestrian if they're also weaving their way across the road with earphones on. So if the vehicles on both lanes pull over to the left, we have a nice wide gap down the centre. The only remaining hazard now is the bollard. So please, when you pull over and stop, try not to be parallel to a bollard. We probably won't be able to squeeze through the gap.

If an ambulance is overtaking, it's usually better to come to a complete stop. If you continue cruising along, you can often create a closing gap, one that we won't be able to pass through. Our vehicles are big and heavy, so it takes us a lot longer to pick up speed again. And we prefer to take it a steady pace, especially if we have a sick patient on board. When we're caring for a very poorly person in the back of the ambulance, it may be necessary for us to stand up or move around to provide treatment or even carry out resuscitation while we're travelling. In this situation, stopping suddenly could prove devastating.

Please don't wind down your window and casually motion us to drive past. This is a particularly male habit and it's quite irritating. We don't need your permission to drive along the road. And it's fair to say that we'd probably have driven past you even if you hadn't wound down your window and gestured to us to do so.

Having your sense of hearing removed from the equation means you need to be more visually alert to the road around you. So if you must drive with your music at deafening levels, it makes sense to check your mirrors more frequently.

The other distraction that can slow us down is using your mobile. We often end up sitting behind someone driving while on their mobile. They seem to be blissfully unaware of the obstruction they're causing us even though they're just a few feet away from the blue lights and sirens. They're also oblivious to the fact that they have every other road user glaring at them in frustration and disbelief while they chatter away. It's illegal. But people continue to do it.

On our patch there is a junction which has four lanes for traffic. I can almost guarantee that as we approach, all the other cars approaching the junction will spread themselves out, blocking every available lane. It's a bit like a game of Tetris or Connect Four. The lights always change to red at this point, so the car drivers then become reluctant to move, leaving us stuck behind a row of cars with no clear lane to go through. But here's the strange bit: all the cars decide to move forward one space. Then they all realise that the other cars have also moved forward one space. So they all move forward one more space. It continues in this fashion until someone takes the initiative and pulls to the side in front of one of the other cars – so we can then continue on our merry way.

Pedestrians, take care too: please don't run in front of us if we're driving along a road with lights and sirens on. That may sound like a very obvious thing to say. But trust me, people do this. It's quite curious: there can be miles of empty road behind us, yet some people still appear compelled, as if by a force from above, to make the dash across the tarmac before we've driven across, rather than after. Not knowing if they are going to make it without tripping over, we are forced to brake in anticipation, only to have to pick up speed afterwards. I've seen parents do this pushing baby buggies in front of them. One woman even had three little children in tow as she took their lives in her hands and dragged them across the road behind her. They were terrified. So was I!

On the same note, when we're roaring along the road, please don't pull out in front of us in your car. I'm convinced that some drivers look and see us approaching as they sit at the junction. They then wait until they can see the whites of our eyes before deciding to pull out on to the road in front of us after all, a bit like a game of chicken. After this hare-brained manoeuvre they immediately indicate and pull over to the side of the road to get out of the way, almost as if they're being helpful. This is too little, too late. Don't pull out in front of us in the first place, please. Because if we do hit the side of your car, you and your passengers will come off a lot worse than us. We may also be a little bit grumpy when you then expect us to jump out and treat you for your injuries. So just be a bit more patient and wait until the road becomes clear behind us before you pull out. You know it makes sense.

ACKNOWLEDGEMENTS

Firstly and most importantly, I would like to give special thanks to my family and friends, my wonderful husband Steve and my three gorgeous children Benn, James and Katie; you make me happy and proud everyday.

I would also like to acknowledge all my fantastic friends and colleagues in the London Ambulance Service, particularly those who have worked with me at HQ and on the Croydon (centre of the universe!) complex. Between you, you have kept me sane, supplied me with tea, chocolate and – only when off duty, of course – wine! You have supported me when required (you know who you are) and at times made me laugh so hard my sides ached and tears streamed down my face. You are the best bunch of people anyone could ever work with, I have absolutely no doubt about that.

I would also like to say thanks to all the marvellous staff at Croydon University Hospital. There were nurses and doctors who became role models to me in my early years as a student nurse. They inspired me to want to be more like them and they continue to influence the way I work today. Some of these people have been there since I started my journey, but I would also like to mention the new friends and colleagues who are there for me now. These good people patiently put up with my incessant questioning and support me in my endeavour to be as good a practitioner as I can be.

Big thanks as well to the staff at the Faculty of Health Care Science at St George's, University of London, where I worked briefly and where I continue to study.

This edition of the book is dedicated to Steve Wright. Steve and I joined the London Ambulance Service on the same day in April 1994. Together with the rest of our group we completed our basic training at Bromley Training Centre. When we finally got to put on the much coveted uniform (which back then was the bright green boiler suits) we had such fun playing at being paramedics, re-enacting all sorts of improbable scenarios. We even had a group song, 'You are my sunshine' which we sang to our long-suffering tutors on our last day in school.

At last, we were let loose on the unsuspecting public as Steve and I were placed together for a six week period of consolidation on a front-line ambulance working from St Helier ambulance station. Steve was always the consummate gentleman and looked after me well during this time. We loved every minute of it and laughed so much as we helped each other muddle through those early days.

Steve's career with the ambulance service was full and varied. He never stopped looking for new challenges.

However, sadly in August 2009 Steve passed away after a short illness. His funeral was a *standing room only* affair with motorcycle out-riders, a piper, and even the helicopter medical service flying past to pay her respects.

Whenever anyone speaks of Steve, they always mention his love for his family, his professionalism, lust for life and pride in his job. They also remember his ever-present beaming smile. Steve was the epitome of what a paramedic should be and I am grateful that his family has kindly allowed me to pay tribute to him by dedicating this edition of my book to his memory.

Best wishes to all the other front-line staff up and down the country. This includes the other ambulance services, voluntary services, fire brigade and especially the police for coming so quickly to rescue us when we call for 'urgent police assistance'. It's good to know that you are there!

Lastly, thanks to the patients who trust us enough to let us into their lives when they are at their most vulnerable and who at times inspire us as they deal with whatever life is throwing at them.

Lysa Walder, London, 2020

AUTHOR NOTE

Heartfelt thanks for taking the time to read my book. If reading my collection of stories here has piqued your interest in becoming a paramedic, I would suggest you read it from the beginning one more time and give yourself a darned good talking to. If, despite these measures, you are still of the same opinion, then you could do worse than to have a wander over to my Facebook group 'So you want to be a paramedic': **https://www.facebook.com/groups/Paramedicwanabees/**

There you will find a lot of information about how to get started on your chosen career path. You will be able to talk to likeminded others who are thinking of applying to study Paramedic Science. There is plenty of advice from existing paramedic students, experienced and specialist paramedics and retired paramedics from all over the UK and beyond. I started this group because, although I originally wrote my book as a memoir for my children and my friends, my unintended audience was perhaps, not surprisingly, young people who were thinking of taking up paramedicine.

Coincidentally, at about the same time, changes were afoot, and paramedicine was evolving into a degree-entry career. Luckily for me, those young people had the propensity to undertake online searches. When 'Paramedic' was entered in the search bar – my book, originally entitled *999; True Stories of my Life as a Paramedic* would feature. Subsequently, I began receiving an increasing volume of correspondence from these young people and, in order to respond to them in a supportive and efficient, manner I started the group.

How could I have known that in 2020 the group would have grown so dramatically? There are now 20,000 members. I have the support of a wonderful admin team to help things run smoothly; Charlotte, Bob, Jon, Neil and Michael do sterling work ensuring there are no online fisticuffs, career ending comments, inappropriate advertising or unsolicited porn spamming slipping in through the back door. If, as a result of reading this book you end up joining our ranks, I hope you enjoy the group – and do please make yourself known!

ABOUT THE AUTHOR

London-born Lysa left school to join a travelling circus, working in Europe for four years as an acrobat and ring-mistress. In a slight change of direction, she qualified first as a nurse then as a paramedic and finally an emergency care practitioner, spending over twenty years saving lives in the UK. She still works shifts in a London Urgent Care centre but is at her happiest when spending time in her old farmhouse nestled deep in the chestnut forests, in the mountains of Tuscany, nursing a glass of prosecco and, finally, living the dream.